Measuring Abundance

Measuring Abundance

Methods for the Estimation of Population Size and Species Richness

Graham J. G. Upton

DATA IN THE WILD SERIES

Pelagic Publishing | www.pelagicpublishing.com

Published by Pelagic Publishing
PO Box 874
Exeter
EX3 9BR
UK

www.pelagicpublishing.com

Measuring Abundance:
Methods for the Estimation of Population Size and Species Richness

ISBN 978-1-78427-232-6 (Hbk)
ISBN 978-1-78427-231-9 (Pbk)
ISBN 978-1-78427-233-3 (ePub)
ISBN 978-1-78427-234-0 (ePDF)

A CIP record for this book is available from the British Library

Cover image: Monarch butterfly (*Danaus plexippus*) migration (iStock/Jodi Jacobson)

Contents

Part III. Mobile individuals

Part IV. Species

Preface

This book aims to bring together, for the first time, descriptions of all the most widely used methods for assessing the sizes of populations of living organisms. The papers referenced come from more than 100 different journals that cover many disciplines. However, both for their ubiquity, and for the number of papers cited, two journals stand out: *Biometrics* and *Ecology*. Together they indicate that the subject of this book might be termed *quantitative ecology*.

Wherever possible, examples are used to illustrate the method being described. The methods selected are either those currently used or the earlier methods that underlay them. In a few cases I have suggested adjustments that appear to improve accuracy.

The techniques and problems associated with the measurement of plant cover, are rather different from those used for assessing the amount of timber in a forest, and are very different from those used for counting birds or fishes. My hope is that specialists working with one type of organism, may chance across a procedure, currently used in a different context, that they can adapt to their own purposes.

The descriptions of the many methods contained in this book are necessarily brief, and there will always be much more that could be written; to cover that deficiency there is a recommended reading section at the back, giving details of specialist books that constitute essential reading for the methods described.

Computer programmes are referenced where appropriate. My preference is for programmes based on R (because they are free), but reference is also made to other widely used programmes. Methods for the assessment of the size of mobile populations are particularly complex. As an example, the online manual for the programme *MARK* (which deals with capture-recapture data) has more than 1000 pages.

Following a brief synopsis of relevant statistical methods in Part I, Part II addresses methods for stationary items. Questions addressed here include: the numbers of standing or fallen trees in a forest; the amount of timber in a forest; the amount of plant cover in a field, and the amount of coral in a coral bank. The methods in this section are relatively easy to describe and use.

It is no surprise that assessing the numbers of moving objects is much more challenging, both to describe, and to carry out. Part III includes examples of the estimation of the numbers of reptiles (skinks), mammals (grizzly bears, marmots), amphibians (frogs), fish (darters), crustaceans (lobsters, crabs), and birds (ovenbirds). Most examples include computer code, though the analyses here would constitute no more than the preliminary stages of a proper analysis of the data.

Part IV is concerned with the many aspects of species richness and diversity.

There are a few sections marked with an asterisk. These are sections that may be read by the curious, but can be ignored without affecting the understanding of the remainder.

About 40 years ago I was co-author (with Bernard Fingleton) of the two volumes of *Spatial Data Analysis by Example*. One review welcomed the publication of the second of those volumes by looking forward to a third volume. We had nothing in mind at the time, but the current volume might have fulfilled that role, since the spatial arrangement of objects has a direct effect on how easy it is to count them.

Graham Upton
Wivenhoe, Essex
June 2020

Acknowledgements

I am very grateful to the following for the help they provided: Richard Barnes (elephant dung), Emery Boose (mapping trees), Nathalie Butt (plotting tree data), Rick Camp (bird data and its provenance), Richard Chandler (*unmarked*), Robert Colwell (Costa Rican ant data), Kevin Darras (audio bird counts), Rocio Duchesne and Ken Tape (Alaskan shrubs), Murray Efford (for advice on spatial capture-recapture), Gregory Gilbert (for permission to use the Californian tree data and for his advice concerning modern methods for mapping trees), Geoff Heard (Australian tree frog data), Klaus Hennenberg (for advice on plotless density estimators), Leanne Hepburn (reef fish protocols), Renske Hijbeek (for supplying the mangrove data and giving permission for its reproduction), Judith Lang and Kenneth Marks (for advice on the coral data they provided), Jeff Laake (advice on mark-recapture), Tom Matthews (ISAR and the Gambin model), Brett McClintock (mark-resight), Trent McDonald (*mra*), David Morrison (nested quadrats), Louis-Paul Rivest (*Rcapture*), Robert van Woesik (SFD), Robert Whittaker (octaves).

I am particularly grateful to Eric Rexstad for his patience and thoroughness in dealing with my barrage of inquiries while learning to use the *Distance* package.

Finally, it is a pleasure to thank Chris Reed and Nigel Massen for their assistance with the preparation of this manuscript.

G. J. G. U.

Part I

Background

Readers with any statistical training should probably move directly to another part of the book. This first part presents the barest possible introduction to the statistical terms that arise subsequently. A reader without statistical training should initially simply skim through this part, returning when the need arises.

1. Statistical ideas

Manly and McDonald (1996) stated that 'Statistical methods play a pivotal role in the process of gathering information to enable many of today's important conservation problems to be solved.' Bonar, Fehmi, and Mercado-Silva (2011) wrote 'Just as a business executive needs the services of a good lawyer and a good accountant, a biologist needs a good statistician.' This book therefore starts with a crash course introducing the statistical terms and methods that will be applied later in the book. This is intended as no more than an *aide memoire* rather than a statistics textbook.

1.1   Sampling

For key species it may be possible to conduct a census with the aim of counting nearly every individual. For example, the 2018 tiger census in India, required 44,000 field staff, 600,000 human-days, and 523,000 km of foot surveys with 26,800 camera trap locations. The results were 35 million photographs of wildlife, of which 76,651 were of tigers. Comparison of markings suggested that 2461 tigers had been photographed. Combining these observed numbers with DNA analysis, and using some of the methods discussed later in this book, led to an estimate of 2967 tigers.[1]

Although it would appear that a census of an entire population must be more accurate than a sample Bonham (2013) noted that 'sample-based data may be more reliable than a 100% inventory. This follows from the fact that samples are often taken with greater care than can be used in a complete census because more expertise can be used in sampling.' While this comment does not apply to the Indian tiger census, which employed a lot of expertise, and used statistical sampling techniques, it does emphasize the potential accuracy that can be obtained using samples.

Consider the modest task of determining the number of oak trees in a large wood. One could methodically walk back and forth through the wood, counting oak trees and marking them to avoid double counting. If the wood is really large, then this will take a very long time and would cost a lot of money. A more sensible idea would be to use a map, divide the wood into 1000 equal-sized small areas, and then visit a *sample* of 10 of these counting the oaks in each small area. Multiplying the total number of oaks seen by 100 (= 1000/10) will give an estimate of the total number in the wood. Noting the variation in numbers across the 10 small regions will give a good idea of the accuracy of that estimate. The numbers 10 and 1000 can be varied as appears appropriate.

The small areas chosen to be sampled must not be chosen because of any prior knowledge. Ideally, they should also not be close to one another, since neighbouring locations are likely to resemble one another without, necessarily, resembling those at the other end of the wood. Alternative sampling schemes (depending on context) will be discussed throughout the book. Underlying every scheme is the requirement that *the sample should be representative of the population being sampled and should in no way be biased by the sampler.* Often the latter requirement implies a random arrangement of samples, or a random sampling point, determined by the computer. Some discussion of where sampling should take place is given in Section 2.3.

1.2 Sample statistics

Suppose that a sample consists of the n values: $x_1, x_2, ..., x_n$. These might be *measurements*, such as the distances from n sampling points to the nearest trees. They might be *counts*, such as the numbers of organisms in randomly chosen equal-sized regions. Natural questions are 'What is their average value?' and 'How variable are these numbers?' The answers are provided by two statistics: the sample mean and the sample variance.

1.2.1 Sample mean

The sample mean, denoted by $\bar{x}$, is the average of the n sample values:

$$\bar{x} = \frac{1}{n}(x_1 + x_2 + \cdots + x_n) = \frac{1}{n}\sum_{i=1}^{n} x_i. \tag{1.1}$$

An equivalent formula is

$$\bar{x} = \frac{1}{n}\sum_{j=1}^{J} f_j x_j, \tag{1.2}$$

where J is the number of distinct x-values and f_j is the number of occurrences of the value x_j. If n is reasonably large (> 30, say) then the distribution of the values of $\bar{x}$ from repeat samples is likely to be well approximated by a normal distribution (Section 1.3.1).

Example 1.1: Californian Douglas Firs

The left-hand diagram of Figure 1.1 shows the positions of Douglas Firs (*Pseudotsuga menziesii*) in a 40 m × 40 m corner of the 200 m × 300 m UC Santa Cruz Forest Ecology Research Plot which forms part of the coastal forest in the Santa Cruz mountains of California. The data[2] were collected between December 2006 and September 2007 and refer to alive main stems that have diameters of at least 1 cm at breast height.

The trees are not uniformly spread across the region surveyed, but have a distinct cluster towards the bottom of the region. The figure shows the region subdivided into a hundred 4 m × 4 m plots, with the counts in these regions reported in the right-hand diagram.

The mean number of firs per 4 m × 4 m plot is

$$\bar{x} = \frac{1}{100}(1 + 1 + 0 + 1 + \cdots + 0 + 0 + 0 + 0) = \frac{65}{100} = 0.65.$$

A more convenient procedure begins by summarizing the data in Table 1.1. The mean can now be calculated using Equation (1.2):

$$\frac{1}{100}\{(63 \times 0) + (27 \times 1) + 2 + (5 \times 3) + (2 \times 4) + 5 + 8\} = 0.65.$$

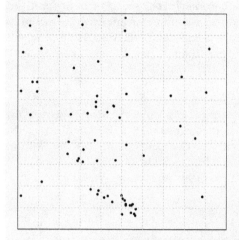

1	1	0	1	0	2	0	1	0	0
1	1	0	0	0	1	0	0	0	1
0	0	1	1	0	0	0	1	0	0
4	0	0	1	0	1	0	1	0	1
1	0	1	3	4	0	0	0	0	0
0	0	1	1	0	0	0	1	1	0
0	0	3	3	1	1	1	0	0	0
0	1	0	0	0	0	0	0	0	0
1	0	0	3	3	8	0	0	1	0
0	0	0	0	0	5	0	0	0	0

Figure 1.1 The left-hand diagram shows the positions of 65 Douglas Firs in one 40 m × 40 m corner of a Californian research plot. The plot is subdivided into a grid of 100 squares of side 4 m. The right-hand diagram reports the counts in each of these 100 squares.

Table 1.1 The numbers of Douglas firs in the hundred 4 m × 4 m quadrats shown in Figure 1.1.

Number of Douglas firs	0	1	2	3	4	5	6	7	8
Number of 4 m × 4 m quadrats	63	27	1	5	2	1	0	0	1

1.2.2 The median, quartiles and inter-quartile range.

Suppose there are 101 distinct sample values labelled $x_1, x_2, ..., x_{101}$ with $x_1 < x_2 < ... < x_{101}$. The *median* is the central value, x_{51}: there are 50 smaller values and there are 50 larger values.

The *lower quartile* is x_{26}: there are 25 smaller values and 75 larger values. Correspondingly, the *upper quartile* is x_{76}: there are 75 smaller values and 25 larger values. The difference in the values of the two quartiles is termed the *inter-quartile range*.

In practice, the numbers of observations and their values are unlikely to be so convenient. But it will always be possible to arrange observations in order of magnitude and identify values that approximate the formal definitions.

1.2.3 Box-whisker plots

A box-whisker plot (also known as a *box plot*) is a way of illustrating the spread of a set of data by using the values of the quartiles and the median. The central box is bounded by the quartiles, with the position of the median indicated either by a point or by a bold line within the box.

The whiskers extend from the quartiles (the edges of the box) towards the more extreme values. Depending on the computer programme used, these either extend all the way to the most extreme values, or they may extend by some multiple of the inter-quartile range, with more extreme values being separately indicated.

Example 1.2: Californian Douglas Firs (cont.)

The Douglas Fir data is far removed from the idealized 101 observations discussed in introducing the quartiles, since so many observations have the same value. The result is Figure 1.2.

In this case, after arranging the data in ascending numerical order, all of the minimum value, the lower quartile, and the median, have the value 0. The upper quartile is 1 and the values of 3 or more are indicated as *outliers*. More usual box-whisker plots will be found in Chapters 4 and 8.

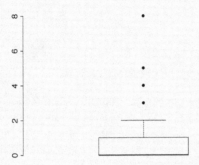

Quadrat counts of Douglas firs

Figure 1.2 The rather unusual box-whisker plot for the Douglas fir data of Table 1.1.

1.2.4 Sample variance

The variability of the n sample values is measured by the sample variance, s^2:

$$s^2 = \frac{1}{n-1}\left\{\sum_{i=1}^{n} x_i^2 - \frac{1}{n}\left(\sum_{i=1}^{n} x_i\right)^2\right\}, \tag{1.3}$$

or, equivalently, by

$$s^2 = \frac{1}{n-1}\left\{\sum_{j=1}^{J} f_j x_j^2 - \frac{1}{n}\left(\sum_{j=1}^{J} f_j x_j\right)^2\right\}. \tag{1.4}$$

Example 1.3: Californian Douglas Firs (cont.)

Using Table 1.1, $\sum x^2 = 27 + 4 + 45 + 32 + 25 + 64 = 197$, giving

$$s^2 = \frac{1}{99}\left(197 - \frac{65^2}{100}\right) = 1.563.$$

In this case, the sample variance is more than twice the size of the sample mean. This suggests that the firs are not distributed at random (see Section 1.4.2 below). This is principally a result of the cluster of firs towards the bottom of Figure 1.1.

1.2.5 Sample standard deviation

This is s, the square root of the variance. The units of s are the same as those of the original data. It is used in the construction of confidence intervals (Section 1.6.2).

1.2.6 Coefficient of variation

Large numbers will usually differ from one another by larger amounts than will small numbers. The coefficient of variation (often referred to as the cv) takes into account the magnitudes of the numbers involved by scaling the standard deviation by the mean:

$$cv = \frac{s}{\bar{x}}.$$

1.3 Common continuous distributions

A *continuous distribution* is appropriate when the quantity being measured is not confined to a specific set of values such as 0, 1, 2, ..., but may take any value in some specified range. It applies to measurements or averages, as opposed to counts.

The curve that illustrates how likely to occur are the possible values, is called the *probability density function*, which is often shortened to *pdf* and written as f(x). A related function is the *distribution function* F(x), which is the probability of obtaining a value of x or less.

1.3.1 Normal distribution

The normal distribution has a shape that depends on the values of the two quantities μ and σ^2, which may be collectively referred to as the *parameters* of the distribution. The distribution is symmetric about the *mode* (the most likely value), μ, which is therefore also both the mean and the *median*. The quantity σ^2 is the variance of the distribution.

About 95% of the values of a normal distribution lie within 2σ of the mean (see Figure 1.3).[3] Although every normal distribution has a theoretical range from $-\infty$ to ∞, about 99.8% lie within 3σ of the mean. This implies that if μ is much greater than 3σ, then the possibility of obtaining a negative value can be ignored.

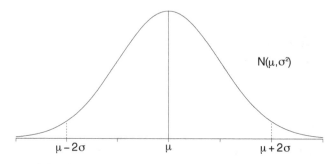

Figure 1.3 A normal distribution centred on μ. About 95% of values lie between $\mu - 2\sigma$ and $\mu + 2\sigma$.

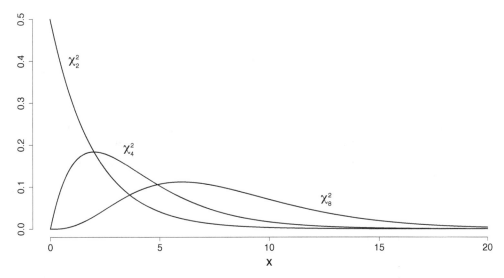

Figure 1.4 Chi-squared distributions with degrees of freedom equal to 2, 4, or 8.

1.3.2 Chi-squared distribution

This is a continuous distribution that can take any value between 0 and ∞. The shape of the distribution is determined by the value of the single parameter ν, which is referred to as the *degrees of freedom* of the distribution. The distribution has mean ν and variance 2ν. Three examples are illustrated in Figure 1.4.

1.4 Common discrete probability distributions

These are distributions that are appropriate when the quantities being measured are counts.

1.4.1 Bernoulli distribution

This simple distribution has just two values: 0 and 1. With the probability of the outcome 1 denoted by p, the distribution has mean p and variance $p(1 - p)$. The Bernoulli distribution plays its part in presence/absence models.

1.4.2 Poisson distribution

When every point in space (or time, or space-time) is equally likely to contain an event, then the events constitute a *Poisson process*. In our case an 'event' means that an organism of interest (a plant, say) is present at that location. The events are said to occur at random.

Suppose that plants occur at random in a region with a density of μ per unit area. In that case the numbers of plants in equal-area subregions are observations from a Poisson distribution. If the subregions have unit area, then the probability of a randomly chosen subregion containing exactly k plants is

$$P(\mu; k) = \frac{\mu^k e^{-\mu}}{k!}$$

where e is the exponential function, and the quantity $k!$, which is referred to as k *factorial*,[4] is given by

$$k! = k(k-1)(k-2)\cdots 1.$$

The mean number of plants in a subregion of area A is μA. Since, for a Poisson process, the variance of the number of plants would also be μA, a comparison of the sample mean, $\bar{x}$ with the sample variance, s^2, provides an indication of whether the plant distribution may be random. If $s^2 \ll \bar{x}$, then this would suggest a very ordered plant arrangement, such as trees in an orchard. If $s^2 \gg \bar{x}$, then this would indicate that the plants occur in clumps.

Example 1.4: Random counts

Figure 1.5 illustrates a Poisson process. The plant positions were generated using pairs of random numbers. A summary of the counts in 100 subregions is given in Table 1.2.

With 100 plants scattered over 100 subregions, the sample mean is 1. The sample variance is given by

$$s^2 = \frac{1}{99}\left\{(0+36+80+72) - \frac{1}{100}(100)^2\right\} = \frac{88}{99} = 0.89.$$

As anticipated, since the plant pattern is genuinely random, the sample mean and sample variance are approximately equal. A formal test is presented in Section 1.8.

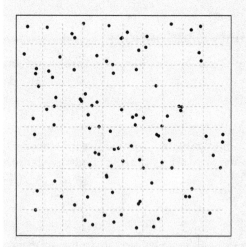

Figure 1.5 The left-hand diagram illustrates the positions of 100 randomly positioned points. The right-hand diagram reports the counts in the 100 subregions.

Table 1.2 A summary table of the numbers of plants in the 100 subregions of Figure 1.5.

Number of plants	0	1	2	3
Count	36	36	20	8

1.4.3 Binomial distribution

The binomial distribution is appropriate for situations with just two outcomes (e.g. 'Success' and 'Failure'). The form of the distribution is determined by the two parameters n, the number of trials, and p, the probability of a success (assumed to be constant across all trials). The probability that there are exactly r successes among the n trials is $P(n, p; r)$ given by

$$P(n, p; r) = \frac{n!}{r!(n-r)!} p^r (1-p)^{n-r} \qquad r = 0, 1, \ldots, n, \qquad (1.5)$$

with 0! defined to be equal to 1. The distribution has mean np and variance $np(1-p)$. Note that the case $n = 1$ corresponds to the Bernoulli distribution (Section 1.4.1).

In cases where there are more than two possible outcome, the probabilities refer to all the possible outcomes and the distribution is called the *multinomial distribution*.

Example 1.5: Random counts (cont.)

The left-hand diagram of Figure 1.6 repeats the counts previously presented, but groups them into 25 sets of four counts. Defining a success as being a count greater than zero, the bottom-right set of four counts includes two successes (1 and 2) and two failures (zeroes). The observed numbers of successes are reported in the right-hand diagram of Figure 1.6.

In 36 of the 100 subregions there are no plants. The estimated probability of a success is therefore $\hat{p} = (100 - 36)/100 = 0.64$. With $n = 4$, and substituting $\hat{p}$ for p in Equation (1.5), Table 1.3 compares the estimated proportions for the outcomes with those observed. In Section 1.8 it will be shown that the differences between the observed and estimated numbers is no more than would be expected by chance.

1	1	2	1	3	1	1	1	2	0
1	1	0	1	2	1	1	0	2	0
2	2	1	0	1	1	0	0	0	0
1	2	2	3	0	0	1	1	0	0
1	2	0	2	1	3	2	3	0	0
1	0	0	1	1	1	2	1	0	3
0	0	0	3	3	2	2	1	0	2
0	1	0	0	2	0	1	0	1	0
1	1	1	1	1	2	0	0	2	0
1	0	0	2	3	1	2	0	0	1

4	3	4	3	2
4	3	2	2	0
3	2	4	4	1
1	1	3	3	2
3	3	4	1	2

Figure 1.6 The left-hand diagram shows the counts in 100 subregions. The right-hand diagram reports the number of non-zero counts within each group of four counts.

Table 1.3 A summary of the numbers given in the right-hand diagram of Figure 1.6 compared to the corresponding figures for a binomial distribution with $p = 0.64$.

Number of successes	0	1	2	3	4
Observed number	1	4	6	8	6
Estimated number (using $p = 0.64$)	0.4	3.0	8.0	9.4	4.2

1.4.4 Negative binomial distribution

The assumption of a Poisson process is mathematically convenient. It underpins many of the distance methods discussed in Chapter 4. In practice, however, there are usually clusters of individuals. With clusters present, the variance of the counts in equal-area regions can be much greater than their mean. In such cases, a negative binomial distribution may provide a better description of the observed counts.

For the negative binomial distribution the probability that exactly k plants are observed, is given by

$$P(r, p; k) = \frac{(k + r - 1)!}{k!(r - 1)!} p^r (1 - p)^k \qquad\qquad k = 0, 1, 2, \ldots. \tag{1.6}$$

Expressed in terms of the distribution's mean, μ, and variance, σ^2, the parameters p and r satisfy the following relations:

$$p = \mu/\sigma^2 \qquad \text{and} \qquad r = \mu^2/(\sigma^2 - \mu). \tag{1.7}$$

Substitution of the sample mean $\bar{x}$ for μ, and the sample variance s^2 for σ^2, gives the so-called *method of moments* estimates as

$$\tilde{p} = \bar{x}/s^2 \qquad \text{and} \qquad \tilde{r} = \bar{x}^2/(s^2 - \bar{x}). \tag{1.8}$$

Example 1.6: Californian Douglas Firs (cont.)

For the 100 small plots illustrated in Figure 1.1 the sample mean and variance were $\bar{x} = 0.65$ and $s^2 = 1.56$. The sample variance is much greater than the sample mean, suggesting that a negative binomial distribution may prove useful. The method of moments estimates are $\tilde{p} = 0.65/1.56 = 0.416$ and $\tilde{r} = 0.463$.

A comparison of the observed counts with those from the negative binomial distribution with parameters $\tilde{r}$ and $\tilde{p}$ are given in Table 1.4. An alternative approach, that often provides a better fit, is maximum likelihood estimation (see Section 1.6.4). The expected values resulting from the maximum likelihood estimates are given as the last line in the table. Neither set of expected values is convincingly superior.

Table 1.4 A comparison of the observed frequencies of Douglas Firs with the estimates obtained using a negative binomial distribution with parameter values being estimated by either the method of moments or the method of maximum likelihood (Section 1.6.4).

Number of Douglas Firs	0	1	2	3	4 or more
Observed number	63	27	1	5	4
Estimated number (method of moments)	66.6	18.0	7.7	3.7	4.0
Estimated number (maximum likelihood)	64.2	20.3	8.4	3.8	3.4

1.5 Compound Poisson distributions

When a population containing many species is repeatedly sampled, the number of individuals belonging to any particular species will vary from sample to sample according to a Poisson distribution. Suppose that, for species j, the mean number per sample is μ_j. If μ_j is regarded as an observation from a continuous distribution with probability density function $f(\mu)$, then P_k, the probability that there are k individuals in the sample, is given by the *compound Poisson distribution*:

$$P_k \propto \int_0^\infty \frac{\lambda^k e^{-k}}{k!} f(\mu) d\mu, \qquad k = 1, 2, 3, \ldots, \qquad (1.9)$$

with the proportionality introduced to take account of the fact that the number of species contributing a zero count cannot be measured.

1.5.1 Log-series distribution

In Fisher, Corbet, and Williams (1943), Fisher's choice of probability density function for the Poisson mean led to the log-series distribution, for which

$$P_k = \frac{\alpha}{kS} \left(\frac{n}{n + \alpha} \right)^k, \qquad k = 1, 2, 3, \ldots. \qquad (1.10)$$

The procedure for estimating the parameter α will be described in Section 10.5.

1.5.2 Poisson-lognormal distribution

Analysing patterns of observed frequencies in samples, Bulmer (1974) suggested that the distribution of the logarithm of the Poisson mean might be a normal distribution. The result is the Poisson-lognormal distribution:

$$P_k = \frac{1}{k! \sigma \sqrt{2\pi}} \int_0^\infty \lambda^{k-1} e^{-\lambda} e^{(\ln(\lambda) - \mu)^2 / 2\sigma^2} d\lambda, \qquad (1.11)$$

where μ and σ^2 are the parameters of the normal distribution.

1.6   Estimation and inference

The purpose of a sample, which consists of relatively few observations, is to deduce the properties of the unmeasured much larger population. As the size of a random sample is increased, so the sample mean, $\bar{x}$, will become more reliable as an estimate of the population mean, μ, as a result of a reduction in the standard error (see below). At the same time, the sample variance, s^2, will become more reliable as an estimate of the population variance, σ^2.

1.6.1   Standard error

The standard error (commonly referred to as the s. e.) is the square root of the variance of some quantity of interest. It is closely related to the sample standard deviation, s. As an example, the standard error of the sample mean, $\bar{x}$, is

$$\sqrt{\mathrm{Var}(\bar{x})} = s/\sqrt{n},$$

where n is the sample size, and s^2 is the sample variance. As the sample size increases, so the standard error reduces, implying less variable values for $\bar{x}$, and increased precision in our knowledge concerning the population mean.

For many situations the normal distribution is relevant, so that on about 95% of occasions the population value lies in the interval (sample value ±2 standard errors), and on about 99.8% of occasions the population value lies in the interval (sample value ±3 standard errors). This is expressed formally as a confidence interval.

1.6.2   Confidence interval

An $\alpha\%$ confidence interval for a population mean is an interval calculated from sample values using a formula that guarantees that $\alpha\%$ of the intervals so calculated will include the true value. Of course, this also means that $(100 - \alpha)\%$ will not include the true value! The interval takes the form:

$$\left(\bar{x} - c\frac{s}{\sqrt{n}}, \quad \bar{x} + c\frac{s}{\sqrt{n}} \right). \tag{1.12}$$

To achieve a given value of α requires larger values of c for smaller sample sizes. For a fixed sample size, larger values of c lead to larger values for α. However, 100% confidence can only be achieved with $c = \infty$. If $n \geq 20$ then the choice $c = 2$ will give an approximate 95% confidence interval. Computer output often includes confidence intervals for quantities of interest.

Example 1.7: Californian Douglas Firs (cont.)

For the $n = 100$ firs, previous calculations gave the summary values the values $\bar{x} = 0.65$, and $s^2 = 1.563$. The standard error is $\sqrt{1.563/100} = 0.125$. Taking $c = 2$, the approximate 95% confidence interval for the population mean is (0.40, 0.90).

1.6.3 Bootstrap interval

Equation (1.12) provides a convenient guide that can be calculated without the need for a computer. However, for small values of n, it relies on assumptions about the population that may not be valid. By contrast, the bootstrap approach makes heavy use of the computer, but is always valid.

Suppose that a sample of n observations is taken and denote the first observation by x_1, the second by x_2, and so on. The bootstrap assumption is that *the value of any new observation will be equal to one of x_1, x_2, ..., x_n with each value being equally probable*. With this assumption one can answer questions about the properties of future samples from the population.

The procedure is to draw further samples, each of size n, with replacement, from this hypothetical distribution. Each new sample is created using random numbers in the range 1 to n. For example, if the first five random numbers are 11, 8, 3, 11, and 4, then the first five observations in the next 'sample' will have values equal to those of x_{11}, x_8, x_3, x_{11}, and x_4, respectively.

Notice that, by chance, some random numbers may occur more than once, while other numbers will not occur at all. For each new sample, the characteristic of interest (for example, the mean) can be calculated. The range of values obtained for that characteristic gives an indication of the uncertainty in its value.

Suppose 999 further samples are generated, so that there are 1000 values of the characteristic of interest (including the value actually observed). An approximate 95% bootstrap interval is

$$(v_{25}, v_{975}),$$

where v_i is the ith largest of the 1000 values. If required, greater precision can be obtained by increasing the number of new samples.

Example 1.8: Californian Douglas Firs (cont.)

The 100 values in the original sample are not distinct, but this does not affect the bootstrap approach, which treats x_1 as the number of Douglas Firs in the first subregion, x_2 as the number of Douglas Firs in the second subregion, and so forth. Table 1.5 summarizes the first three resamples.

Figure 1.7 is a *histogram* (a diagram in which the areas of rectangles represent counts) that illustrates the wide range of the means obtained from resampling the Douglas Fir data. The values ranged from 0.33 to 1.07. Arranging the values in order, the bootstrap 95% confidence interval was found to be (0.42, 0.92), which is in excellent agreement with the (0.40, 0.90) given using Equation (1.12) (with $c = 2$). Note that each new set of 999 resamples may lead to a slightly different interval.

Table 1.5 The numbers in the first three resamples of the hundred 4 m × 4 m Douglas Fir counts summarized in Table 1.1.

Number of Douglas Firs	0	1	2	3	4	5	8	Mean
Original data	63	27	1	5	2	1	1	0.65
First resample	56	31	1	7	2	3	0	0.77
Second resample	62	30	2	3	1	1	1	0.60
Third resample	60	30	0	6	1	2	1	0.70

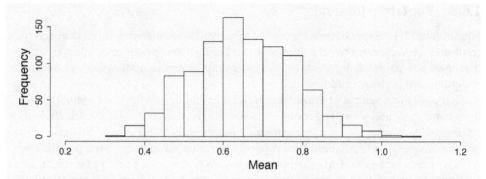

Figure 1.7 Histogram showing the distribution of the means of 999 bootstrap resamples of the data summarized in Table 1.1.

1.6.4 The likelihood function and maximum likelihood estimates

For a random sample $x_1, x_2, ..., x_n$ of observations on X, the *likelihood* is the product of the probabilities of their occurrence:

$$L = \{P(X = x_1) \times P(X = x_2) \times \cdots \times P(X = x_n)\}. \tag{1.13}$$

Often these probabilities will be functions of one or more unknown parameters. The values of those parameters that maximize L are called the *maximum likelihood* estimates.

In some cases there will be no need to use the computer to maximize L as simple formulae exist (for example, for a Poisson distribution with parameter μ, the maximum likelihood estimate of μ is just the sample mean, $\bar{x}$).

When the likelihood is a function of a single unknown parameter, a 95% confidence interval for that parameter consists of all values that lead to a value of the log(likelihood) that is within 1.92 of the maximum value.[5] Finding these values requires a simple loop programme on the computer.

When the likelihood is a function of two unknown parameters, the joint 95% confidence region for the two parameters consists of all pairs of values that lead to a value of the log(likelihood) that lies within 3.00 of the maximum value.[6] In this case, a *contour plot* of the likelihood will demonstrate the interdependence of the two-parameter estimates.

Example 1.9: Californian Douglas Firs (cont.)

The full counts of the Douglas Fir data were given in Table 1.1. Assuming a negative binomial distribution, the likelihood for these data is given by

$$L = \{p^r\}^{63} \times \{rp^r(1-p)\}^{27} \times \cdots \times \left\{ \frac{(r+7)!}{8!(r-1)!}p^r(1-p)^8 \right\}^1.$$

The estimates arising from maximizing L are $\hat{r} = 0.6154$ and $\hat{p} = 0.4863$. The corresponding expected values were given in Table 1.4.

Figure 1.8 shows contours of the likelihood for these data. The uncertainty in the estimation of the parameter values is apparent: this is not a sharp mountain but a rather flat topped ridge. The location of the maximum is indicated by the filled dot. The hollow dot indicates the values of the method of moments estimator found previously. To convey the uncertainty in the estimates, they might be reported as $\hat{r} \approx 0.6$ and $\hat{p} \approx 0.5$.

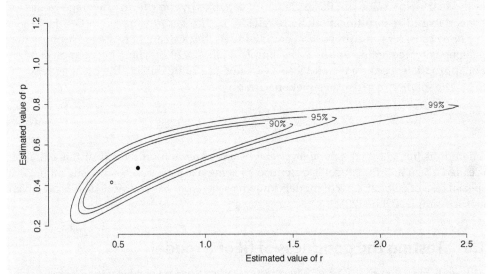

Figure 1.8 Contour plot of the likelihood surface for the Douglas Fir data. The maximum likelihood estimates are indicated by the filled dot. The method of moments estimates are indicated by the empty circle which lies well inside the joint 90% confidence region indicated by the inner contour.

1.7   Types of model

Many models are extremely complex, involving information on many variables, and defying simple description. For example, when using capture-recapture methods to estimate the number of mobile animals in a study area, the model requires estimation of the probability that an animal is present, together with estimation of the probability that an animal present is detected. These probabilities may depend on a variety of characteristics, such as age and gender. This section briefly describes two simple types of model that may be incorporated in more sophisticated models.

Linear models are principally used to relate a variable of interest, y, to one or more explanatory variables. The simplest form is the *linear regression model* typified by

$$y = \beta_0 + \beta_1 x, \tag{1.14}$$

where x is the value of an explanatory variable, and the βs are constants whose values are estimated using a sample of pairs of values of x and y.

When there are several relevant explanatory variables, the model is extended in a natural fashion:

$$y = \beta_0 + \beta_1 x_1 + \cdots + \beta_k x_k. \tag{1.15}$$

Here there are k *explanatory variables* and this is called a *multiple regression model*.

The early literature on these linear models assumed that the variable of interest had a normal distribution. However, modern computer programmes permit the assumption of many different distributions, and the resulting models are referred to as *generalized linear models*, or simply GLMs. Still more recently, these models have been extended, by replacing individual 'point' values for the explanatory variables by some form of average values for the variables. The resulting models are called *generalized additive models* or GAMs.

Now consider p, the proportion of individuals that are caught in a trap. Suppose that p depends on the temperature, x. The simple replacement of y by p in equation (1.14) is not appropriate, because it could lead to a value of p lying outside the possible range (0 to 1). The solution is to use an equation such as:

$$\ln\left(\frac{p}{1-p}\right) = \alpha + \beta x, \tag{1.16}$$

where ln is the so-called *natural logarithm*.[7] The expression on the left of this equation is called a *logit* and the model is described as being a *logistic model*. A logistic model is a special case of a wider class of models known as *log-linear models*. More details are given in, for example, Upton (2016).

1.8   Testing the goodness of fit of a model

Although there are many specialized tests available for particular situations, a useful general-purpose test is *Pearson's goodness of fit test*. The test requires the calculation of X^2 given by

$$X^2 = \sum_{j=1}^{J} \frac{(O_j - E_j)^2}{E_j}, \tag{1.17}$$

where O_j and E_j are, respectively, the observed and estimated counts in category j, and J is the number of categories. If the model under test is correct, then the value obtained for X^2 will be typical of a chi-squared distribution having $J - 1 - P$ degrees of freedom, where P is the number of parameters with values estimated from the data. For the chi-squared approximation to be reliable, all the E values should be at least 3. If the value obtained for X^2 does not lie in the upper tail of the distribution, then the model under test may be judged acceptable.

An alternative to X^2 is G^2, the *likelihood ratio goodness of fit statistic*, or *deviance*, which is given by

$$G^2 = 2 \sum_{j=1}^{J} O_j \ln\left(\frac{O_j}{E_j}\right). \tag{1.18}$$

In most cases the value of G^2 is close to the value of X^2 and, as with X^2, if the model under test is correct, then the value of G^2 will be typical of a chi-squared distribution having $J - 1 - P$ degrees of freedom.

With a hand calculator X^2 is easier to calculate, but, since G^2 has some theoretical advantages, when a computer programme is doing the hard work, it is usually the value of G^2 that is given in its output.

Example 1.10: Californian Douglas Firs (cont.)

The observed and estimated frequencies using a negative binomial distribution were given in Table 1.4. In this case $J = 5$ (the number of categories reported) and two parameters were estimated from the data. The reference distribution is therefore a chi-squared distribution with $5 - 1 - 2 = 2$ degrees of freedom. The goodness of fit statistics are given by

$$X^2 = \frac{3.6^2}{66.6} + \frac{9^2}{18.0} + \frac{6.7^2}{7.7} + \frac{1.3^2}{3.7} + 0 = 10.86,$$

$$G^2 = 2\left\{63\ln\left(\frac{63}{66.6}\right) + 27\ln\left(\frac{27}{18.0}\right) + \cdots + 4\ln\left(\frac{4}{64.0}\right)\right\} = 13.82.$$

The probability of obtaining a value of 10.86 or more, from a chi-squared distribution with 2 degrees of freedom, is about 0.004.[8] The probability of a value greater than 13.82 is about 0.001. It seems that a negative binomial distribution (with the parameter values estimated using the method of moments) provides a very unlikely description of the data.

Example 1.11: Random counts (cont.)

Table 1.3 reported the observed and fitted counts of successes based on genuinely random data. In this case, the estimated number for the category '0 successes' was very small (0.4), so, to improve the accuracy of the X^2 test, that category is combined with the next to give the category '≤ 1 success'. The estimated frequency for the combined category is 3.41 and the observed frequency is 5. Thus

$$X^2 = \frac{1.59^2}{3.41} + \frac{1.96^2}{7.96} + \frac{1.44^2}{9.44} + \frac{1.81^2}{4.19} = 2.22.$$

After combining the categories, $J = 4$. The corresponding value for G^2 is 2.10. The data were used to estimate the value of one parameter (p), so the reference distribution is a chi-squared distribution with $4 - 1 - 1 = 2$ degrees of freedom. These values are close to the average value for a chi-squared distribution with two degrees of freedom, so the hypothesis that the points were scattered randomly would be accepted.

1.9 AIC and related measures

This book is concerned with inferring the size of a population from information provided by samples from that population. It is usually true that, with two competing models that describe the sampled data, it is the more complicated model that provides the more accurate description. As an extreme example, suppose that a description is required for the following set of measurements, which have been taken at regular intervals:

$$0, 10.4, 19.8, 32.0, 39.9, 50.3.$$

One possible description is 'the first observation is 0, the second is 10.4, the third is 19.8, the fourth is 32.0, the fifth is 39.9, and the sixth is 50.3'. This description is 100% accurate, but it is very long-winded and could not be described as a summary. By contrast the statement that 'each observation is about 10 more than its predecessor with the sequence starting at zero', while not a perfect description, is usefully concise and effectively describes the values.

AIC (short for the *Akaike Information Criterion*) (AIC) is a measure, introduced by the Japanese statistician Akaike in 1973, that balances model complexity against goodness of fit. AIC values are reported by many computer programmes. The value reported for a particular model is not in itself important; what matters is how large that value is compared to the values reported for alternative models. The best model (*providing that it makes sense to the investigator*) is the model with the smallest AIC value. If it were simply a matter of comparing AIC values, then the investigator would be replaced by the computer! A discussion of this point is provided by Mac Nally et al. (2017) who ask the question 'is the "best" model any good?'

For logistic or log-linear models (Section 1.7) the value of AIC is equal to a constant (that depends on underlying distributional assumptions) plus the value of the deviance G^2 (Equation (1.18)) reduced by $2d$, where d is the degrees of freedom associated with the G^2 value. Adding extra information into a model (for example, whether a creature is a male or a female), will never increase the value of G^2, but it will decrease d. If the information is not sufficiently valuable, then AIC will increase. Some programmes report the value of AICc, which incorporates a correction for sample size. The difference between AIC and AICc is usually very small.

The *Bayesian Information Criterion* (BIC) works in the same way as AIC, but associates a greater penalty to the loss of a degree of freedom. The two measures have subtly different motivations: AIC seeks to select that model, from those available, that most closely resembles the true model (which will be governed by a myriad of unmeasured considerations and will *not* be among those considered), whereas BIC assumes that the correct model *is* among those on offer and seeks to identify that optimal model.

Since, in the context of the measurement of abundance, none of the models considered are likely to provide a perfect description, AIC should be the measure used.

1.10　Quantile-quantile plots

In distance sampling (Chapter 8) it is necessary to fit curves to a histogram of frequencies. A difficulty with histograms is that they require data to be grouped into ranges of values: different groupings give rise to different histograms and, potentially, to different impressions concerning how well the curve fits the data. An alternative, that avoids grouping the data, is the *quantile-quantile plot*, which is often called the *q-q plot*.

Suppose that the observed values are numbered in increasing order, so that $x_1 \le x_2 \le \ldots$ x_n. So (roughly) a proportion $1/n$ of the observed values are less than or equal to x_1, while $2/n$ are less than or equal to x_2, and so forth. In a q-q plot these proportions[9] are plotted against the corresponding theoretical tail probabilities $F(x_1)$, $F(x_2)$, ..., $F(x_n)$, where $F(x)$ is the theoretical distribution function (Section 1.3) under test.

Figure 1.9 shows two examples of q-q plots. For a good-fitting model the plots lie close to the line of equality. In a poor-fitting model there will be noticeable divergences.

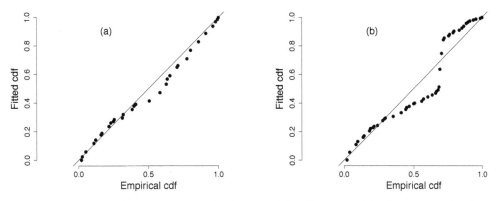

Figure 1.9 Two q-q plots. (a) An example of a model that fits acceptably. (b) An example of an unacceptable fit.

1.10.1 Cramér-von Mises test

This test was suggested by the Swede, Cramér, in 1928, and, independently, by the Austrian, von Mises, in 1931. Denoting the observed distribution function by $\widehat{F(x)}$ and the corresponding theoretical function by $F(x)$, it collects together all the discrepancies in the q-q plot into a single statistic W^2, given by

$$W^2 = \frac{1}{12n} + \sum_{i=1}^{n} \left\{ \widehat{F(x_i)} - F(x_i) \right\}^2 . \tag{1.19}$$

Large values of W^2 suggest that the theoretical distribution is incorrect.

Part II

Stationary individuals

It might seem easy to count stationary objects, but the length of this part of the book is an indication that life is not that simple. The questions that arise are where to count, what to count, and how to count. If the intention is to count all the individuals in a given region, then rules are required for individuals that overlap the region's boundary. The counting procedure must be both practical and reliable. Much of the discussion in the next few chapters centres on finding efficient methods to achieve objectives. Before continuing, the reader might ponder on the problems associated with counting daisies in a playing field, and then contrast the chosen method with counting oak trees in a wood. One size does not fit all.

2. Quadrats and transects

The term 'quadrat' originally referred to a square wooden sampling frame (typically of area 1 m^2). It now refers to a region of any shape and size (but usually rectangular or square) that will be used to take repeated samples from a population of interest. Quadrats might be used to estimate species abundance (richness), to determine which species are present, or to determine the amount of ground cover (or canopy cover) attributable to particular species. A narrow rectangular quadrat may be described as a *transect*.

In this book it will be assumed that there is a random element to the placement of a quadrat. However, there is an alternative procedure in which the region examined (known as a *relevé*) is carefully chosen to be fully representative of the wider area of interest. That choice must be made by someone with a deep knowledge of the general region. Workers using relevés are particularly interested in the interactions between species. Their aim might be to create a taxonomy of vegetation types. This branch of study has been termed *phytosociology*.

2.1 What shape quadrats?

Whatever shape quadrat is used, there must be clear rules relating to plants that overlap the quadrat boundary. For a rectangular or square quadrat, a typical rule would be that the plant is counted if it overlaps the north or east boundaries, but not if it overlaps the south or west boundaries. This avoids double counting or overestimation of species abundance. For further discussion of problems caused by the edge see Section 2.4.

With small quadrats it will be useful to take a photograph in case of doubt when collating results.

2.1.1 Advantages of rectangular quadrats

- In Section 9.3.1 it is shown that, on average, more species are likely to be found in rectangular plots than square plots of the same area.
- It is easy to keep track of which parts of the quadrat have already been searched. With a square quadrat it might be necessary to mark out subdivisions to avoid double counting or missed sections.
- Narrow rectangular quadrats also called *belt transects*, *strip transects* (or simply *transects*) might be easily monitored by a single observer travelling down the plot centre (for example, a diver monitoring the ocean bottom), or by two observers travelling in tandem down either edge of the plot.[1]

- Narrow rectangular quadrats result when *remote underwater video* (RUV) technology is used to monitor cover on the sea bed. A comprehensive review of alternative approaches is provided by Mallet and Pelletier (2014).
- Narrow rectangular quadrats are convenient if it is necessary to avoid entering the quadrat (in order not to tread on, or otherwise damage, the individuals of interest).

---

❶ *Advice on data collection*

For rare plants, Elzinga et al. (1998) suggest using a quadrat with a width of 0.25 m or 0.5 m.

2.1.2 Advantages of square quadrats

- Square quadrats are the natural choice when a grid of contiguous quadrats is used.
- Boundary overlap is less of a problem with a square quadrat than it would be for a rectangular quadrat of the same area, because of the smaller perimeter of a square. For example, a square 1 m² quadrat has a perimeter of 4 m, whereas a rectangular 2 m × 0.5 m quadrat has a 5 m perimeter.
- Squares are a natural choice when a square grid is already in existence for other purposes. For example, in the UK, there is a well-defined 1 km square National Grid created for mapping purposes. In the USA, the Public Land Survey System of the Bureau of Land Management uses 1-mile-square regions.

2.1.3 Advantages of circular quadrats

- Circular quadrats are easily defined with a central pole and a rope in clear ground. Bibby et al. (2000) recommended circular quadrats in the context of burrow-nesting seabirds.
- Kershaw et al. (2016) provide a useful discussion of the relative merits of circular and rectangular (or square) plots. They observe that since circular plots have the smallest possible perimeter for a given area, their use minimizes boundary problems.

2.2 How many quadrats?

Any increase in the number of quadrats sampled will result in an increase in accuracy. However, since resources are finite, a practical approach is to use the smallest number that gives the desired accuracy. Suppose that there are n quadrats, each of area a, being used to estimate the total number of individuals in a region with area A. Let the sample mean number of individuals be denoted by $\bar{x}$, with the sample variance being s^2. The estimate of the number of individuals for the entire region, $\widehat{N}$, is given by

$$\widehat{N} = \frac{A}{a}\bar{x}. \tag{2.1}$$

Comparison of the values of $\bar{x}$ and s^2 gives an idea of the spatial pattern of the locations of the species concerned. If s^2 is much less than $\bar{x}$ then the individuals are rather regularly placed in the study region. If s^2 is much greater than $\bar{x}$ then the individuals occur in clusters. The intermediate case is a random pattern (which typically appears visually to be a mix of regular and clustered components).

An approximate 95% confidence interval (see Section 1.6.2) for the number of individuals in the region is

$$\left(\widehat{N} - 2\frac{As}{a\sqrt{n}}, \quad \widehat{N} + 2\frac{As}{a\sqrt{n}} \right). \tag{2.2}$$

To gain an idea of the implications of these formulae, suppose that the objects of interest are randomly located in space with an average of λ per unit area. The randomness assumption implies that the objects form what Matheron (1989) termed a *Poisson forest*. In this case, with quadrats of unit area, both the mean and the variance of the observed counts will be λ. Since $a = 1$, the approximate 95% confidence interval for the number of individuals in an area A will then be

$$\left(\lambda A - 2A\sqrt{\frac{\lambda}{n}}, \quad \lambda A + 2A\sqrt{\frac{\lambda}{n}} \right). \tag{2.3}$$

Expressing the width of the interval as a proportion of the value being measured gives

$$4\sqrt{\frac{1}{n\lambda}}.$$

If this is to be no greater than 25%, then simple maths shows that the required number of quadrats is $256/\lambda$. Thus eight quadrats would be sufficient if they were large enough to contain on average 32 individuals, but far more quadrats would be required if the average per quadrat was much less.

With regularly spaced individuals, fewer quadrats are required than would be the case for a Poisson forest, but with clustered individuals far more sampling may be necessary. In the absence of clear information on the spatial distribution, a pilot study will be useful.

To see the effect of the spatial distribution of the sampled individuals, consider the three cases illustrated in Figure 2.1. In each case there are 100 individuals within a region of 100 square units being sampled by the same eight quadrats of size 1 square unit.

The extreme case (a) has a rectangular grid of individuals, with the grid spacing exactly matching the quadrat size, so that every quadrat contains exactly one individual. The resulting estimate of 100 individuals in the sampled region is exactly correct and without doubt. Just one quadrat would have been enough!

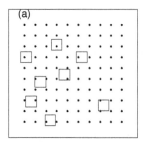 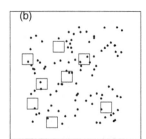 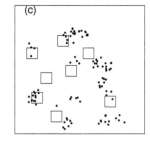

Figure 2.1 Three equal-sized regions each containing 100 individuals of interest. The three patterns are (a) regular, (b) random, and (c) clustered.

Case (b) is an example of a random pattern (a Poisson forest). The eight quadrats have counts of 0, 1, 1, 1, 1, 2, 2, and 2 giving $\bar{x} = 1.25$, $s^2 = 0.5$, and an approximate 95% confidence interval as (75, 175). The confidence interval includes the correct value, but it is a wide one indicating that more quadrats are needed to give a useful interval.

Case (c) illustrates a highly clustered pattern with quadrat counts of 0, 0, 0, 0, 1, 2, 3, and 9, giving $\bar{x} = 1.875$, $s^2 = 9.55$. Note that s^2 is much greater than $\bar{x}$. In this case Equation (2.2) gives the approximate 95% confidence interval as (–31, 406). The lower limit is clearly nonsense: it points to the need to gather more information. Increasing the number of quadrats to 32 does provide a feasible interval of (63, 212), but this is still very wide, and yet more sampling would be required before a usefully precise estimate could be obtained.

Example 2.1: Californian trees (cont.)

The counts for the trees of Figure 1.1 were summarized in Table 1.1. For these counts $\bar{x} = 0.65$, and $s^2 = 1.54$, so that the approximate 95% confidence interval for the number of Douglas firs in the entire plot is:

$$\frac{200 \times 300}{16} \left(0.65 \pm 2 \sqrt{\frac{1.54}{100}} \right) = 2437.5 \pm 931.6 = (1506, 3369).$$

The estimate $\hat{N} = 2437.5$ is reassuringly close to the true number of trees (2183), though the confidence interval is so wide that it would probably be felt that further sampling was required.

---

❶ Advice on data collection

Historically, trees were mapped using measuring tapes and compasses, though the latter were somewhat imprecise, and the resulting maps might contain serious errors. As a result, Rohlf and Archie (1978) suggested methods based only on distances. When these were distances between two trees, any error in one tree position would lead to an error in the position deduced for the next tree. Boose, Boose, and Lezberg (1998) proposed an alternative distance-only method, that minimized the possibility of error amplification.

A more straightforward procedure, that eliminates the possibility of accumulating errors, begins by setting out a carefully checked rectangular grid of marker posts at fixed distances across the study area. Subsequently, for each tree, its overall location can be determined by determining the direction of, and distance to, a nearby marker post (taking account of the diameter of the tree concerned). Typically, the measuring instruments are lasers and sighting compasses. This was the procedure used with the Californian survey plot.

Currently, mapping often involves the use of GPS. However, the accuracy of GPS varies across the Earth's surface, depending on the number of satellites for which there is a line-of-sight view. The accuracy has increased over time, with an increase in the number of satellites, and improvements in software and hardware. In June 2016, according to the US government, 95% of positions were reported as being measured correctly to within 2 m horizontally and 3 m vertically. In some locations, the GPS information can be augmented by other systems, to give much improved accuracy. Rayburn, Schiffers, and Schupp (2011) claimed accuracy to within 2 cm when locating shrubs by using a survey-grade base unit, together with a rover unit mounted on a fixed height pole equipped with a bubble level.

2.3   Quadrat placement

In the previous example, the 100 quadrats were arranged in a contiguous 10 × 10 square. In practice, if the idea is to survey a much larger area, then this would be a poor choice, since neighbouring locations tend to be similar to one another, and may give a poor guide to the nature of distant locations within the area of interest.

With scattered quadrats some form of random allocation is required to avoid conscious or unconscious bias. However, centring quadrats at randomly selected points, as illustrated in Figure 2.2 (a), may result in overlapping quadrats and an uneven coverage.

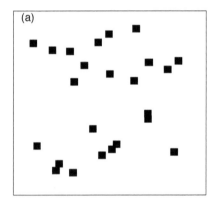
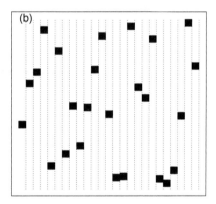
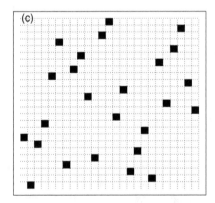
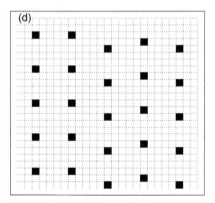

Figure 2.2 (a) A random arrangement of quadrats. (b) An arrangement having one quadrat randomly positioned in each column. (c) An arrangement in which each row and each column contains exactly one quadrat. (d) A systematic arrangement with fixed gaps along rows and columns.

Overlapping quadrats are easily avoided by using some form of stratification. Figure 2.2 (b) illustrates an allocation in which each column contains one randomly placed quadrat. However, there still appears to be some clustering. This is reduced in Figure 2.2 (c) where both rows and columns contain a single quadrat.[2]

Figure 2.2 (d) illustrates *systematic sampling*. In this case a value of x is randomly chosen in the range 1 to 5, with the next sets of five values having x co-ordinates at intervals of 5 from the first. For each set, the same procedure is used to determine the y co-ordinates. This process guarantees that every quadrat has an equal chance of selection, with an even coverage for the entire study region.

In a large-scale survey there may well be different terrain types or ecosystems (for example, urban, farmland, woodland, etc.). To ensure that all are represented in an appropriate way, random (or systematic) sampling should take place within each. This is called *stratified sampling*.

2.3.1 Generalized random-tessellation stratified design (GRTS)

A problem with the idealized arrangements illustrated in Figure 2.2 (c) and (d) is that they may not work! For example, the chosen quadrats may miss important features of the study region; they may be inaccessible or they may have inappropriate habitat. Stevens and Olsen (2004) developed a procedure that provides a spatially balanced random sample while nevertheless avoiding these problems. The stages of their procedure are:

1. Establish a labelled grid of quadrats across the study region.
2. Arrange the labels in order along a 'line', giving each an appropriate 'length'. Suppose the length of the 'line' is N.
3. Suppose n quadrats are required. Let N/n be denoted by k and let m be a value chosen at random in the range 1 to k.
4. The quadrats chosen are those corresponding to the locations (along the line) m, $m + k$, $m + 2k$, etc.

Figure 2.3 illustrates the first stages in the labelling of quadrats for GRTS. The study region is divided into four sections; each section is divided into four subsections; each

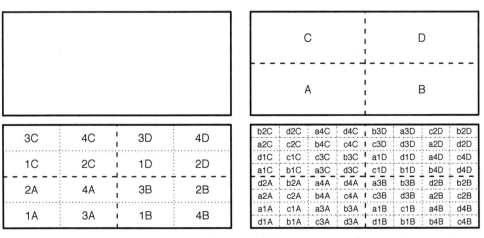

Figure 2.3 The study region is divided into 4 sections. Each section is broken into 4 subsections. Each subsection is again split. This continues until quadrats of the required size are obtained.

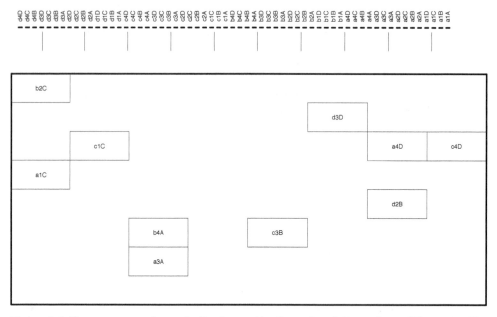

Figure 2.4 The upper part shows the line formed by the ordered 61 quadrats of Figure 2.3. The lower part shows the ten selected quadrats.

subsection is further divided. The process continues until the smallest quadrat size is deemed appropriate for the application. The quadrats are then arranged in order using *reverse hierarchical ordering*. For the case illustrated that order is:

d4D, d4C, d4B, d4A, d3D, d3C, ..., d1B, d1A, c4D, c4C, ..., a1A.

In the example there are $N = 64$ quadrats. Suppose that three central quadrats (c1D, a3B, and d4A) are unusable (length 0), and that the remainder are equally important (length 1). The resulting 'line' is illustrated in the upper part of Figure 2.4.

Suppose we want to choose $n = 10$ quadrats. Thus $k = (64 − 3)/10 = 6.1$. Suppose the randomly chosen start is $m = 4.31$. The chosen quadrats are those corresponding to the points 4.31, 10.41, 16.51, ..., 59.21. The process and the outcome is illustrated in Figure 2.4.

In this example all 61 usable quadrats were given equal weight. On some occasions, however, there will be a minority of quadrats that correspond to habitats that the investigator is particularly keen to sample. For example, if quadrat c3D is assigned a 'length' 10, rather than 1, then its probability of inclusion will be 10 times that of the other quadrats.

2.4 Forestry sampling

Although there are some specific challenges that face foresters, much of this section has wider application.

2.4.1 Cruising

In forestry, timber stocks may be assessed by examining the trees in a sequence of sample plots, often arranged along a series of parallel paths, in the study region (this is referred

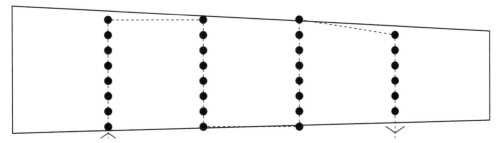

Figure 2.5 Forestry sampling may consist of a 'cruise' with samples taken at regular intervals (here using circular plots) on a series of parallel paths through the forest. Sample plots that overlap the edge need special treatment.

to as *fixed-plot sampling*). Foresters refer to the process of travelling from one plot to the next as *cruising*.

An example cruise is illustrated in Figure 2.5. Note that some sample plots overlap the edge; the subsequent analysis will need to take this into account (forest edges should be sampled since they are likely to have a different composition from forest centres).

2.4.2 Cluster sampling

When sampling forests, a major cost is likely be associated with the journey to a sampling point. Once that point has been reached, it will therefore be cost-effective to take several samples before moving to the next sampling point. The number of subplots, and the distances between them, must be decided without reference to the local terrain. Some examples of possible configurations and routes are shown in Figure 2.6.

According to Yim et al. (2015), configurations (a) and (d) have been used in national forest inventories in South Korea, with (b) being used in Germany. Their study considered the times associated with the stages of the sampling process and they also considered the effects of spatial correlation on the accuracy of estimates. They recommended (d) for future use in South Korean forests. The impact of quadrat placement on apparent species diversity is discussed in Section 9.3.1.

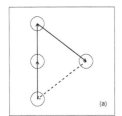
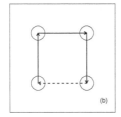
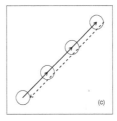
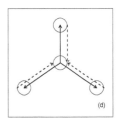

Figure 2.6 Example configurations of forestry subplots, with possible connecting routes (dotted lines).

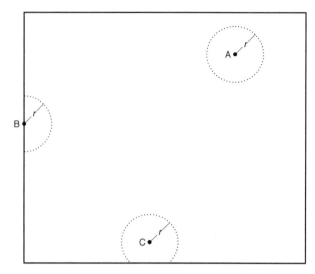

Figure 2.7 A, B, and C are locations in the study region. Circular quadrats of radius r are being used. A would be included in the sample if the quadrat centre lay anywhere within the dotted circle shown. B would be included by any quadrat whose centre lay within the semicircle shown. C is more likely to be included than B, but is less likely than A.

2.4.3 Slopover bias

In Figure 2.7 circular quadrats of radius r are in use. Locations A, B, and C lie within the region of interest. Whether or not they are included in a sample will depend upon the location of the sampling points. A sampling point centred anywhere on, or within, the dotted circle of area πr^2 surrounding A, will result in A being included. For B, however, the relevant region is not a circle, but a semicircle of area $\pi r^2/2$. Thus A is twice as likely to be selected as B. Any location, such as C, that is within r of the edge will have a reduced chance of being selected. In forestry, this bias, which results from quadrats potentially overlapping the edge of the study region, is termed *slopover bias*.

Note that slopover bias exists whether or not any quadrats actually overlap the edge and irrespective of the shape of the quadrat. The bias applies wherever quadrats are used, but may be particularly important in forestry, where the species composition and plant density at the edge may be distinctly different from that in the centre of the study region.

One way of eliminating slopover bias is to allow the quadrat centre to be placed up to a distance r outside the study region (Masuyama, 1954). However, this would not be feasible if, for example, the study region bordered a motorway, or a large body of water.

If the proportion of a quadrat that lies within the study region is known, then one solution is to scale up the plant count. For example, if 3/5 of the quadrat lies inside the study region, then the count made for that quadrat should be multiplied by 5/3. Here are two other methods of eliminating the bias.

2.4.3.1 The mirage method

This method, introduced by Schmid-Haas (1969), can be used when the study region has a straight edge. If a quadrat overlaps the edge, then the overlap is notionally folded back into the study region with plants in the folded area being added to those already

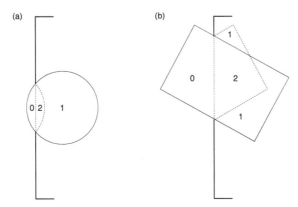

Figure 2.8 The mirage method for correcting slopover bias. (a) With a circular quadrat, plants in the reflected area get weight 2. (b) With a rectangular quadrat having a different orientation from the study region, some plants will get weight 2, but the reflected region may also result in new plants being selected.

counted. For a circular quadrat this will mean that some plants are counted twice. This is illustrated in Figure 2.8 (a).

With a square or rectangular plot, if the quadrat is at an angle to the edge of the plot, then new plants may be counted. This is illustrated in Figure 2.8 (b). Gregoire (1982) demonstrated that the procedure results in unbiased estimates of abundance.

2.4.3.2 The walkthrough method

Ducey, Gove, and Valentine (2004) suggested a variant on the mirage method that treats each near-edge plant individually. Crucially, it does not require a straight edge to the boundary. The decision on whether a plant should be doubly weighted, is based on whether a plant at twice the distance in the same direction, would fall outside the region of interest (double weighting), or inside the study region (single weighting) Examples are shown in Figure 2.9. Valentine et al. (2006) suggest variations of the procedure for use with cluster sampling.

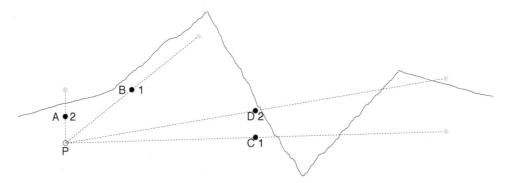

Figure 2.9 The walkthrough method. From sampling point P, the plants at A and D are doubly weighted, whereas those at B and C do not get counted twice because, at twice the distance, the paths from P fall within the study region.

2.4.4 Circular quadrats for sampling coarse woody debris

The term 'coarse woody debris', often shortened to CWD, refers to the fallen tree trunks (or large branches) lying on the floor of a forest. Gove and van Deusen (2011) identify three methods that may be used when a fallen tree straddles the boundary of a circular quadrat of radius R.

2.4.4.1 The stand-up method

For this method the fallen tree is imagined as standing up on the spot where its base currently rests (which may not be where it originally grew). If that spot lies within the quadrat then the entire fallen tree is included in the sample. Otherwise, the tree is ignored.

Suppose sampling is performed using N quadrats of radius r. Denote the number of fallen trees in quadrat j by n_j, with the relevant measurement (e.g. the tree's length, volume, or surface area) for tree i being denoted by y_{ji}. Using only quadrat j, the estimate for the entire region (of area A) being sampled is

$$\widehat{t}_j = \frac{A}{\pi r^2} \sum_{i=1}^{n_j} y_{ji}. \tag{2.4}$$

Combining information from all N quadrats gives an overall estimate as

$$\widehat{t} = \frac{1}{N} \sum_{j=1}^{N} \widehat{t}_j. \tag{2.5}$$

2.4.4.2 The chainsaw method

If the tree trunk is judged to be straddling the plot boundary then the amount included is that which lies within the plot (having hypothetically been cut out of the entire tree using a chainsaw). Gove and van Deusen (2011) describe three possible protocols for judging whether a fallen tree straddles the plot boundary:

1. If any part of the trunk falls within the plot.
2. If any part of the trunk's centre line (from top to bottom, through the core of the trunk) falls within the plot.
3. If any cross section falls within the plot.

Of these, only the first is unbiased (Gove and van Deusen, 2011).

2.4.4.3 The sausage method

If any part of the trunk's centre line (from top to bottom, through the core of the trunk) falls within the plot, then the entire tree is included in the sample. For a trunk to be included, the centre of the quadrat must lie within r of the trunk's centre line. The resulting 'catchment region', illustrated in Figure 2.10, was likened to a sausage by Gove and van Deusen (2011).

Gove and van Deusen (2011) discuss the relative merits of the three methods. They note that the sausage method is likely to measure more trees than the other methods,

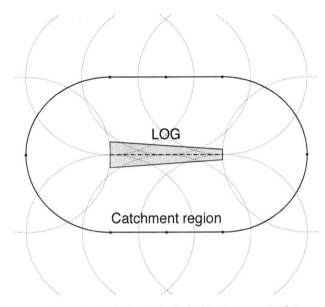

Figure 2.10 The sausage method: the log is included in the sample if the quadrat centre lies within the catchment region. Quadrats centred on the edge of the catchment region touch the centre line of the log.

thereby causing more work, but giving less variable estimates. They note that a fallen tree that has broken into pieces will cause problems when using the stand-up or sausage methods. However, calculating the value of interest (e.g. the volume) is more difficult with the chainsaw method.

An alternative approach to the measurement of CWD is discussed later in Section 5.4. A comprehensive review of alternative methods is provided by Russell et al. (2015).

2.5   Quadrats for estimating frequency

In the present context, frequency is defined as the proportion of equal-sized quadrats that contain the species of interest. As Figure 2.11 indicates, frequency should not be confused with quantity.

Since, when measuring frequency, the exact number of individuals is not required, results are obtained speedily with excellent agreement to be expected between independent observers.

Using a grid of contiguous quadrats of an appropriate size, frequency can be a useful guide to the regions where the species is scarce or abundant. However, if the quadrats are too large, then the species may appear to be omnipresent, while, if the quadrats

A	A	B	B
			B
		B	

Figure 2.11 Species A has a greater frequency (2/3) than species B (1/3), but there is a greater quantity of species B (3 individuals) than species A (2 individuals).

are too small, there will be much unnecessary recording. A good size for a quadrat is therefore one such that between 25% and 75% of quadrats contain the species of interest. If several species are of interest, and some are scarce, then it may be necessary to combine contiguous quadrats, or use nested quadrats of differing sizes (see Section 2.6). Note that comparisons involving frequency can only be valid if they are made using data from quadrats of identical size and shape.

Frequency maps can be a fast and effective means for monitoring change in species occurrence, though there needs to be a clear definition of what is meant by 'occurrence'. For a plant, 'occurrence' might be defined as having roots within the quadrat (so-called *root frequency*); for a tree that might mean having its trunk in the quadrat; for an animal or a bird the equivalent would be that the creature has its home in the quadrat. To avoid double counting, rules are required for organisms that straddle boundaries. For example, an organism may be counted if it straddles a 'south' or 'west' boundary, but not if it straddles the 'north' or 'east' boundaries.

A more liberal definition of occurrence in a quadrat simply requires that some part of the object occurs in the quadrat. In the context of shrubs, this is referred to as *shoot frequency*. For mobile creatures, the equivalent would be that the creature is simply observed in (or passing through) the quadrat. Whatever definition of frequency is used, the critical requirement is that it is clearly specified, and is the same for any comparisons across space and time.

If a species is present in r of the n quadrats, then the frequency is defined as r/n. If the intention is to use these results to estimate the frequency for the entire region sampled, then some idea of the uncertainty in the estimate will be required. Confidence intervals for proportions are not straightforward propositions (see Upton (2016) for details), and the most appropriate approximate 95% confidence interval appears to be that suggested by Agresti and Coull (1998), which is

$$\left(\tilde{p} - 2\sqrt{\frac{1}{n}\tilde{p}(1-\tilde{p})}, \quad \tilde{p} + 2\sqrt{\frac{1}{n}\tilde{p}(1-\tilde{p})} \right) \qquad \text{where} \quad \tilde{p} = \frac{r+2}{n+4}. \qquad (2.6)$$

Example 2.2: Californian trees (cont.)

Figure 2.12 once again shows the positions of the Douglas firs (*Pseudotsuga menziesii*) in the corner of the Californian tree plot. Since the species occurs in each of the 10 m × 10 m squares, quadrats of that size would be ineffective in monitoring the frequency of the species. The smaller 4 m × 4 m quadrats are much more effective: 37% of these include the species.

The approximate 95% confidence interval for the frequency in the entire Research Plot is

$$\frac{39}{104} \pm 2\sqrt{\frac{1}{100}\frac{39}{104}\frac{65}{104}} = 0.375 \pm 0.097 = (0.28, 0.47).$$

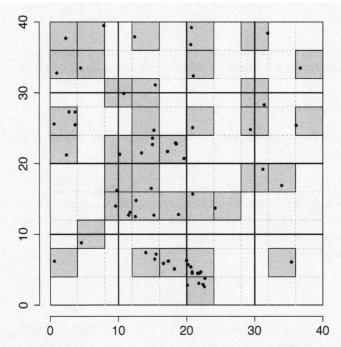

Figure 2.12 The locations of 65 Douglas Firs in one 40 m × 40 m corner of the Californian Research plot. Quadrats of side 10 m are ineffective for showing frequency, since Douglas Firs occur in each. The smaller quadrats, of side 4 m, are more appropriate. Douglas Firs occur in 37% of these.

2.5.1 Estimating abundance from frequency data

Suppose that N individuals are randomly distributed across a region that is subdivided into Q quadrats. The number occurring in a quadrat is therefore an observation from a Poisson distribution with mean N/Q. The probability of a quadrat containing at least one individual is $1 - e^{-N/Q}$. Suppose that q quadrats contain at least one individual. Then q/Q is an estimate of $1 - e^{-N/Q}$. Taking natural logarithms and rearranging gives an estimate of N as

$$-Q \ln(1 - q/Q).$$

He and Gaston (2000) used a similar argument to arrive at the more accurate estimate:

$$\widehat{N} = \frac{\ln(1 - q/Q)}{\ln(1 - 1/Q)}. \tag{2.7}$$

Acknowledging that randomness was uncommon, Yin and He (2014) proposed assessing the departures from randomness, by using a grid of quadrats, and counting the number of occasions on which adjacent quadrats were occupied. If individuals are randomly distributed in an $a \times b$ grid of quadrats so that q quadrats are occupied, then the probability of a pair of quadrats both being occupied is

$$P = \frac{q(q-1)}{ab(ab-1)}.$$

For an $a \times b$ grid of quadrats there are $(2ab - a - b)$ possible neighbours (see Figure 2.13). Thus the expected number of neighbouring pairs (assuming a random distribution) is

$$E = P(2ab - a - b).$$

Denoting the observed number of neighbouring occupied quadrats by O, the ratio I, given by

$$I = O/E,$$

provides an indication of departures from randomness. A value for I less than 1 indicates some degree of clustering, whereas a value greater than 1 indicates some degree of regularity. Yin and He (2014) suggested adjusting the estimate $\widehat{N}$ by taking account of the value of I and using

$$\tilde{N} = \begin{cases} \widehat{N}/I^{\widehat{N}^c} & \text{if } I < 1, \\ \widehat{N}I^{\widehat{N}^c} & \text{if } I > 1, \end{cases} \tag{2.8}$$

where c is a constant with a value close to zero. They suggested evaluating $\tilde{N}$ using both $c = 0.01$ and $c = 0.1$ to give an indication of the uncertainty in the estimate. Note that $\tilde{N}$ is never less than $\widehat{N}$ and that the values chosen for c are based only on empirical evidence.

The following example suggests that the values obtained for N and $\tilde{N}$ should only be regarded as indications of the unobserved abundance.

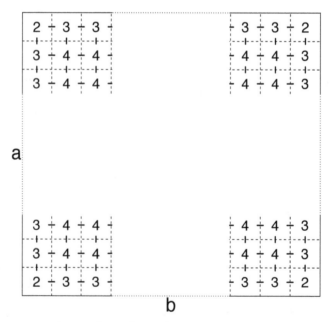

Figure 2.13 Entries are the numbers of neighbouring quadrats. Central quadrats each have four neighbours. Edge quadrats lose a neighbour. Corner quadrats lose 2 neighbours. Since each link joins two quadrats, there are (2ab – a – b) distinct links.

Example 2.3: Californian trees (cont.)

In Figure 2.12, a 10 × 10 grid of 4 m × 4 m quadrats was used, so $a = b = 10$ and there were $2ab - a - b = 180$ neighbouring pairs of quadrats. There are $q = 37$ quadrats containing at least one Douglas Fir. Using Equation (2.7), the estimated number of Firs under the hypothesis of randomness is

$$\widehat{N} = \frac{\ln(1 - 0.37)}{\ln(1 - 0.01)} = 46.0.$$

The value of P is 37 × 36/9900 = 0.1345, which gives $E = 24.2$. The observed number of pairs, O, is 25 which suggests very slight clustering. The ratio $I = O/E$ is found to be 1.0322. Since $I > 1$, the revised estimate of the number of Douglas Firs is given by $\tilde{N} = \widehat{N}I^{\widehat{N}c}$. Using $c = 0.01$ gives $\tilde{N} = 47.5$, while the choice $c = 0.1$ gives the estimate 48.2. Since the true number is 65, for these data the method has fared poorly.

Suppose that the top and bottom halves of the illustrated region differ materially in habitat. With that knowledge each half should be separately examined. The upper half has 21 occupied quadrats with 12 neighbouring pairs giving $\widehat{N} = 27.0$ and the range for $\tilde{N}$ as (33.0, 35.3). The true count was 30. The lower half has 16 occupied quadrats and 11 neighbouring pairs giving $\widehat{N} = 19.1$ and the range for $\tilde{N}$ as (25.4, 27.7). The true count was 35. In both cases there has been an appreciable under-estimate.

2.6 Nested quadrats

2.6.1 Nested quadrats for general use

When several species are of interest, with some being common and others scarce, it will be difficult to find a single quadrat size that works well for all species. One solution, suggested by Whittaker (1977), is to use a design that incorporates a variety of quadrat sizes.

Two possible rectangular designs based on a 20 m × 50 m plot are illustrated in Figure 2.14. The first design, used by the Carolina Vegetation Survey, is derived from that suggested by Peet, Wentworth, and White (1998). The second design, which is used in the USA by the National Institute of Invasive Species Science (NIISS), is the so-called Modified-Whittaker plot recommended by Stohlgren, Falkner, and Schell (1995). The plot should be oriented to maximize the vegetation gradient so as to collect the largest possible variety of species.

Figure 2.15 (a) shows a 100 m^2 design advocated by the Food and Agriculture Organization of the United Nations. Information on different aspects of the ecosystem are collected at different scales (1 m^2 herbs, 25 m^2 for shrubs, and 100 m^2 for trees). Figure 2.15 (b) shows the implementation by Vicharnakorn et al. (2014) of a modified form of the design as part of larger 1600 m^2 quadrats.

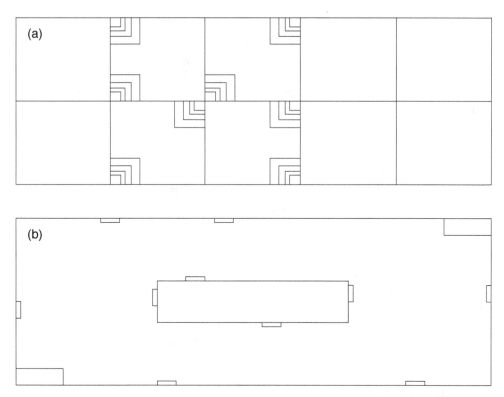

Figure 2.14 Two arrangements for nested subplots within a 20 m × 50 m plot. (a) The design used by the Carolina Vegetation Survey. (b) The design recommended by the National Institute of Invasive Species Science.

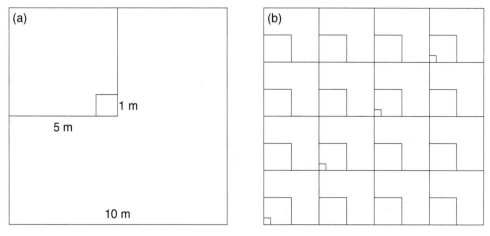

Figure 2.15 (a) An arrangement for nested subplots within a square plot of side 10 m suggested by the Food and Agriculture Organization of the United Nations. (b) An implementation of a modified form of (a) within 1600 m² quadrats used by Vicharnakorn et al. (2014) for the remote sensing of an area bordering Thailand.

2.6.2 Nested quadrats for estimating frequency

Outhred (1984) suggested an alternative approach to the measurement of frequency using concentric square quadrats. The aim is to simultaneously obtain information on rare and common species. The design consists of concentric square quadrats, with areas that roughly double in magnitude, as illustrated in Figure 2.16. Outhred suggested using two measures, that he termed frequency score and importance score. The *frequency score* is defined as the proportion of the regions in which a plant occurs. In the figure, since the plants occur in four of the seven regions, the frequency score is 4/7. The *importance score* is determined by noting the innermost region within which the plant occurs. In the figure the two central regions are devoid of plants, so the innermost plant occurs in the fifth region, counting from the outside. The importance score is therefore 5/7.

Morrison, Le Brocque, and Clarke (1995) carried out an extensive investigation of Outhred's suggestions. They concluded:

> The importance-score method involves no more sampling effort than does standard qualitative (presence-absence) sampling, and it can therefore be used to sample a larger quadrat area than would normally be used for frequency sampling. This makes the method much more cost-effective as a means of estimating abundance, and it allows a greater number of the rarer species to be included in the sampling. The frequency-score method is more time-consuming, but it is capable of detecting more subtle community patterns. This means that it is particularly useful for the study of species-poor communities or where small variations in composition need to be detected.

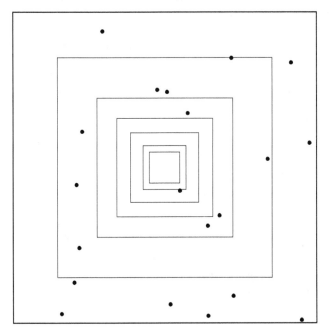

Figure 2.16 A sampling design consisting of seven concentric square quadrats having areas of 1, 2, 5, 10, 20, 50, and 100 units. Twenty randomly placed plants are shown. Since they occur in four of the seven regions the frequency score is 4/7. Since the innermost plant is in the fifth region, starting from the outermost ring, the importance score is 5/7.

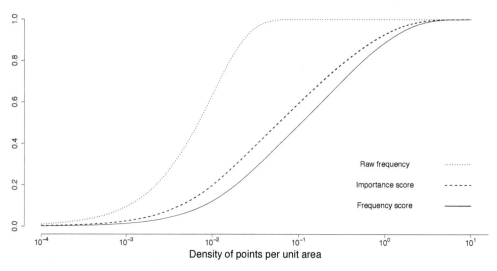

Figure 2.17 The average frequency, importance score, and frequency score for randomly positioned plants shown as a function of the density of the plants. Whereas frequency is linearly related to density over a single order of magnitude, the two scores are linearly related over at least two orders of magnitude, making them better at distinguishing between the relative frequencies of different species.

Figure 2.17 illustrates the improved sensitivity achieved by using these scores for the case of randomly placed points. In the figure, the mean score (or mean frequency) is plotted against the true point density. For this example it is assumed that the seven concentric square quadrats having areas of 1, 2, 5, 10, 20, 50, and 100 units.

All three measures have similarly shaped curves, but, crucially, the range of values where frequency is in the range (0.1, 0.9) is much greater for the new measures. The ratio of the density corresponding to a score of 0.9, to that corresponding to a score of 0.1, is about 20 for frequency, but nearly 140 for the frequency score, and 180 for the importance score. In Figure 2.17 this is shown by the much steeper slope for the midsection of the curve.

---

❶ *Advice on data collection*

'The key thing is to construct stringlines that define the diagonals of the nested squares, with appropriate markers along them, to mark the corners of the squares. We used a central metal pole, with stringlines extending from there to the four corners. The markers defined the corners of squares of 0.25, 0.5, 1, 2, 5, 10, 20, 50, 100, 200, 500 and 1000 sq m. The users then need to identify the edges of each of the squares by eye, imagining the edges joining the equivalent stringline markers. If there are problems, then an extra stringline, placed successively along each edge of each nested square, should deal with it.' (Morrison, pers. comm.)

2.7   Quadrats for estimating cover

Cover (also called *coverage*) is a measure of the influence of an organism. However, while there is a clear idea of what cover means, its definition may vary from one study to another, and its measurement will rarely be precise.

2.7.1   Alternative definitions of plant cover

Fehmi (2010) suggests that three definitions of plant cover are in frequent use, with different names being used by different authors. The definitions (with Fehmi's descriptive names) are:

1. **Aerial cover:** Data refer only to species directly viewable from above. The maximum percentage for any one species is 100%, and the total coverage for detected species cannot exceed 100%. Cover deduced from satellite images is of this type. The total aerial cover may be described as the extent of *foliar cover*, or *crown cover* for the region. In sparsely vegetated areas it may be more appropriately termed *ground cover*.
2. **Species cover:** For each species this is the proportion of the region covered by that species. The maximum for any one species is 100%, but, since species may overlap, the total species coverage may exceed 100%.
3. **Leaf cover:** When point quadrats are used and every distinct overlap of a sampling point is separately counted, the value calculated for a single species may exceed 100%.

While aerial cover may be relatively easy to determine from high quality satellite images, it is not so easy to determine from the ground, where an observer will be viewing tree tops at an angle. The proportion of the sky that is visible to a stationary observer may be termed *canopy closure* (Jennings, Brown, and Sheil, 1999).

The term *canopy cover* has been used with each of the above definitions, and for a variant of aerial cover that uses circumscribing circles or polygons to represent individual plants. This latter definition leads to greater coverage values than those for the aerial cover described above. The UK Forestry Commission uses the term *net canopy cover* for measurements to the drip line of an individual tree. It uses *gross canopy cover* to include the additional regions between the canopies of neighbouring trees together with small glades. To understand any set of coverage values the 'small print' must therefore be studied to determine exactly what has been measured, particularly when comparing data from independent sources.

Basal cover is the proportion of ground area occupied by plant bases (for example, tree trunks). Since this varies little from year to year, it is useful when looking at time series. It is also used in judging the relative importance of the species present.

Example 2.4: Alaskan shrubs

As part of a NASA-funded research project to map changes in the abundance of tall shrubs (heights greater than 0.5 m) in Arctic tundra a field campaign was carried out in 2010 and 2011 on the North Slope of Alaska. Details are given by Duchesne, Chopping, and Tape (2016).

The aim was to match ground truth with available high-resolution panchromatic satellite imagery, in order to arrive at an estimate of the overall abundance of shrubs in that area. This example concentrates on a single 250 m × 250 m site (number 5) bordering the Chandler River.

The data, which are downloadable from the Distributed Active Archive Center for Biogeochemical Dynamics,[3] include crown height (height from ground to topmost branch), crown radius (half the distance from the leftmost branch to the rightmost branch as encountered by the observer), and the spatial co-ordinates determined by GPS. Canopy cover is estimated as the area of a circle of the calculated crown radius.

The sampling protocol consisted of surveying five parallel equi-spaced transects, 5 m wide, spanning the 250 m square plot. Figure 2.18 shows that, given the practical difficulties, the survey team did an excellent job at following the protocol: the dotted lines indicate the apparent positions of the transects (deduced from the recorded shrub positions). The figure illustrates the locations of 99 shrubs (all alders) at 80 distinct locations (in cases where shrubs were very close together a single GPS reading was used for all), with the crown dimensions indicated. Cases of apparently off-transect shrubs can be attributed to unreliable GPS readings.

The figure suggests that the shrub density varies considerably across the plot. The observed canopy cover proportions (truncating radii to 2.5 m for shrubs wider than the 5 m transect) for the five transects are 0.1075, 0.0817, 0.0471, 0.0555, and 0.0297, giving $c = 0.064$, $s^2 = 0.000934$, and an approximate confidence interval for the cover for this plot as (0.037, 0.092).

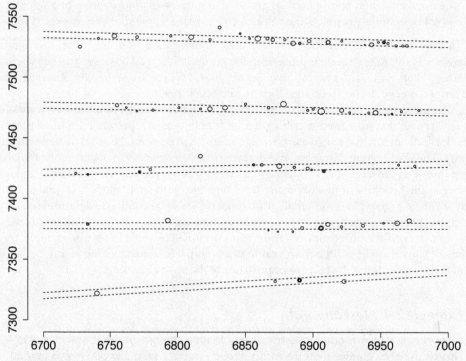

Figure 2.18 The recorded locations of Alders (*Alnus*) in five transects forming part of a 250 m × 250 m site bordering the Chandler River in the North Slope of Alaska. The circles reflect the shrub crown radii with concentric circles indicating cases where a single GPS reading was assigned to several shrubs. The dotted lines indicate the approximate transect paths deduced from the recorded data.

ⓘ *How the data were obtained*

To establish the positions of the shrubs, the observers walked along the centre line of each transect carrying a 5-metre rod. Shrubs were included if they intersected the rod and had their base within the transect. When a shrub was encountered, the shrub was measured and photographs were taken both of the shrub and of the GPS record.

The shrub radius was measured as half the distance from the leftmost branch to the rightmost branch as encountered by the observers. The canopy height was taken to be the height from ground level to the topmost branch.

At the time of the survey the GPS records were known to be occasionally grossly misleading, but generally accurate to within 10 m. Given the expected GPS accuracy, when several shrubs were close together, a single GPS record was used.

2.7.2 Correcting edge effects for cover assessment

When the plants providing cover are relatively large compared to the study region, the rules concerning which plants should be included can make a considerable difference to the cover assessment. Two possible rules are

1. Consider only plants lying wholly in the quadrat.
2. Consider both plants lying wholly within the quadrat and also those (of any size) that overlap the quadrat.

In Figure 2.19, which illustrates these rules, the dotted region indicates the region within which the quadrat centres must lie. For Rule 1 (cases A and B in the figure) larger plants are disadvantaged relative to smaller plants because there is a smaller region within

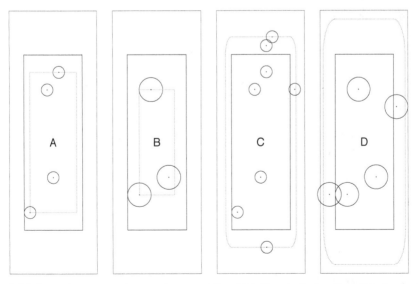

Figure 2.19 In each case quadrat centres must lie within the dotted region. If, to be recorded, plants must lie wholly within a quadrat, then larger plants are disadvantaged (compare cases A and B). If plants overlapping the edge are included, then smaller plants are disadvantaged (compare cases C and D).

which the centres of the large plants can lie. For Rule 2 (cases C and D) the reverse is true, with large plants being easier to detect than small plants.

Correction requires a scaling of the area of the dotted regions in Figure 2.19 to that of the original quadrat. Suppose that the quadrat is rectangular with sides of length a and b and that plant i may be approximated by a circle with diameter d_i. Then plant i should be assigned weight w_i, where, for Rule 1:

$$w_i = \frac{ab}{(a - d_i)(b - d_i)},$$

(2.9)

and, for Rule 2:

$$w_i = \frac{4ab}{4(a + d_i)(b + d_i) - (4 - \pi)d_i^2}.$$

(2.10)

With n circular plants detected in a quadrat the proportion of the quadrat covered by plants is estimated by

$$p = \frac{1}{4ab} \sum_{i=1}^{n} w_i \pi d_i^2.$$

(2.11)

ℹ Advice on data collection

For an irregular shaped plant that is roughly circular (for example, the crown of a tree), an appropriate value for d may be the average of the length and the breadth, where the length is defined as the maximum diameter of the object and the breadth is the length of the diameter at right angles to the length.

Example 2.5: Bahamian coral

Table 2.1 uses data derived from the online database[4] made available by AGRRA (Atlantic and Gulf Rapid Reef Assessment). I am grateful to their contributors and data managers for making their extensive data available. The data in the table refers to a transect, 1 m wide and 10 m long, traversed on 29 April 2011, at Anguilla Cays (part of the Cay Sal Bank in the Bahamas). There were six corals noted with maximum diameters (lengths) of at least 4 cm (the minimum recordable size). The coral measurements and associated cover calculations are given in the table.

The total effective coral cover is 1565 cm² in the study region of 100,000 cm²: thus coral is estimated to cover less than 2% of the total area at this location.

Table 2.1 Corals of at least 4 cm in diameter recorded (using Rule 2) on a 1 m × 10 m Bahamian transect.

Length (cm)	6	8	15	16	18	50	Total
Breadth (at 90° to the length; cm)	3	6	11	12	18	40	
Estimated value for d (cm)	4.5	7	13	14	18	45	
Weight, w	0.9527	0.9282	0.8739	0.8654	0.8330	0.6619	
Weighted cover = $w\pi d^2/4$ (cm²)	15	36	116	133	212	1053	1565

2.7.2.1 Size-frequency distribution (SFD)

Zvuloni et al. (2008) note that 'Many areas of ecological research aim toward characterizing the size-frequency distribution (SFD) of populations to assess change across space and across time.'

If some individuals are large relative to the size of a quadrat, then one of Equations (2.9) and (2.10) should be used to correct the apparent size-frequency distribution for biases due to edge effects.

2.7.3 The visual assessment of cover

Visual assessment is unreliable, since observers will inevitably report round numbers (10%, 20%, ...) or simple fractions (25%, 33%, ...) rather than amounts such as 7% or 26%.[5] Another problem is that the totals of visual assessments of the cover provided by individual species may well exceed 100%. An example of these difficulties follows.

Example 2.6: Dartmoor quadrats

Kent and Coker (1992, Table 3.4) report the assessments of ground cover made by students for 25 quadrats in Dartmoor. Table 2.2 summarizes the totals reported.

The most extreme result was a total of 195%. For that quadrat, according to the students, *Pteridium aquilinum* (a fern) covered nearly the entire quadrat while both *Festuca ovina* (a grass) and *Galium saxatile* (Heath bedstraw) each covered about half the quadrat.

The counts of the cover percentages reported for individual species are illustrated in Figure 2.20. The strong preference for 5%, 10%, and 20% is evident.

Table 2.2 Totals of the cover assessments made by students for 25 Dartmoor quadrats.

Cover total	100–119	120–139	140–159	160–179	180–199
No. of quadrats	5	9	7	2	2

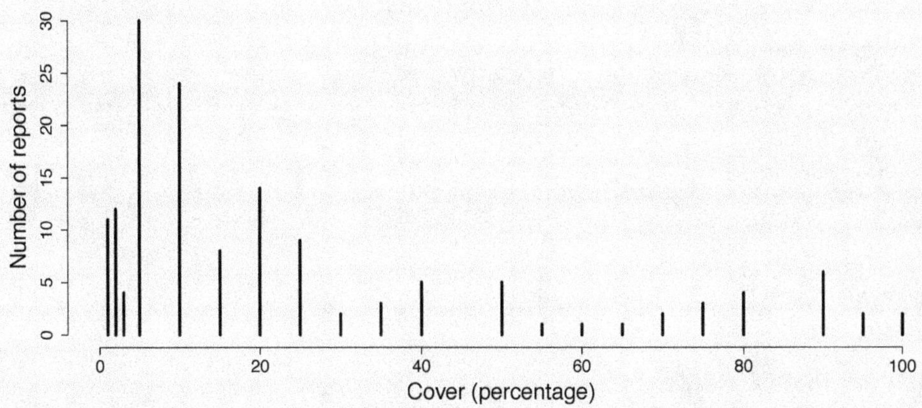

Figure 2.20 The cover percentages reported for the individual species observed in 25 Dartmoor quadrats.

The use of cover scales for plants

In the field, accurate cover measurements are well-nigh impossible, and any attempt at great accuracy would be very time-consuming. For these reasons, in 1885, the Finnish botanist Ragnar Hult suggested using a simple five-category geometric scale, now referred to as the Hult-Sernander-Du Rietz cover scale.

A scale providing more detail for dominant plants is due to Daubenmire (1959), who suggested using a portable rectangular quadrat with sides of lengths 20 cm and 50 cm, coloured so as to make the assessment of coverage using his six-point scale particularly quick and easy.

Another alternative, which places more emphasis on the rarer species, is the New Zealand scale, which is a slight modification of the Braun-Blanquet scale given later in Table 2.5. The three cover scales are summarized in Table 2.3.

Cover abundance is sometimes described using the ACFOR scale, where A, C, F, O, and R, are shorthand for Abundant, Common, Frequent, Occasional, and Rare. The descriptions may be used either informally, or may be guided by a scale such as that given in Table 2.4 (in which case an organism accounting for more than 75% of cover might be described as Dominant).

Braun-Blanquet (1928) and Domin (1928), independently suggested scales that measured plant abundance using a combination of cover and richness. Both scales have subsequently been modified to provide finer details, especially for rarer species, with the modifications being due to Barkman, Doing, and Segal (1964) and Krajina (1933), respectively. These scales are given in Table 2.5, along with the simpler North Carolina scale suggested by Peet, Wentworth, and White (1998).

For the visual estimate of cover to be reasonably accurate, a rectangular quadrat should not be too long (since that would make estimation of a proportion difficult), nor too narrow (since then the proportion of plants overlapping an edge might be large).

Cover may be regarded as a surrogate for biomass, since the direct measurement

Table 2.3 Alternative cover scales.

Hult-Sernander-Du Rietz cover scale								
Scale		1	2	3	4			5
Cover: upper limit (%)		6.25	12.5	25	50			100
Daubenmire cover scale								
Scale	1			2	3	4	5	6
Cover: upper limit (%)	5			25	50	75	95	100
New Zealand cover scale								
Scale	1	2		3	4	5		6
Cover: upper limit (%)	1	5		25	50	75		100

Table 2.4 The ACFOR scale.

Description	Abundant	Common	Frequent	Occasional	Rare
Cover: upper limit (%)	75	50	25	5	1

Table 2.5 Alternative cover-abundance scales.

Braun-Blanquet cover-abundance scale

Scale	r	+	1		2	3	4		5
Cover: upper limit (%)	5	5	5		25	50	75		100
Number of individuals	1	F	N						

Extended Braun-Blanquet cover-abundance scale

Scale	r	+	1	2 m	2a	2b	3	4	5
Cover: upper limit (%)	5	5	5	5	12.5	25	50	75	100
Number of individuals	1–3	F	A	VA					

Domin cover-abundance scale

Scale	1	2	3	4	5	6	7	8	9	10
Cover: upper limit (%)	4	4	4	10	25	33	50	75	90	100
Number of individuals	F	S	M							

Domin-Krajina cover-abundance scale

Scale	+	1	2	3	4	5	6	7	8	9	10
Cover: upper limit (%)	1	1	1	5	10	25	33	50	75	99	≈ 100
Number of individuals	1	F	N								

North Carolina cover-abundance scale

Scale	1	2	3	4	5	6		7	8	9	10
Cover: upper limit (%)		1	2	5	10	25		50	75	95	100
Number of individuals	F										

Key: F = Few; N = Numerous; A = Abundant; VA = Very abundant; S = Several; M = Many

of biomass is time-consuming and, by definition, destroys the plant being measured. Chiarucci et al. (1999) compare results from measurements of cover and biomass; it appears that cover is generally an effective substitute.

Example 2.7: Dartmoor quadrats (cont.)

It might be thought that using a cover scale rather than a visual estimate would greatly reduce accuracy. However, a study by Damgaard (2014) found that 'the rather rough Braun-Blanquet sampling procedure provided cover estimates that were comparable in accuracy' to other sampling methods. Given their similarity, the same would be true for other cover scales. Figure 2.21 demonstrates this for the Daubenmire scale applied to the Dartmoor data where the observed percentages have been replaced by the midpoints of the Daubenmire classes (2.5%, 15%, …, 97.5%).

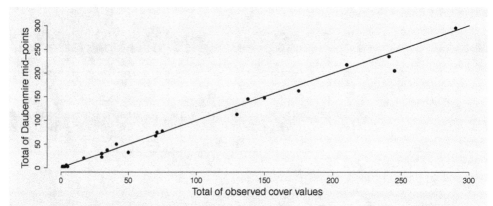

Figure 2.21 The total Dartmoor cover percentages for the 23 commonest species plotted against the estimates that would have been obtained if the cover observations had been recorded using the Daubenmire cover scale.

2.7.4 The SACFORL cover scale

The plant cover scales given in Tables 2.3 to 2.5 took no account of the size of individual organisms since they referred only to total cover. A more sensitive scale is provided by The Marine Biological Association of the UK. Their scale, used since 1990, takes account of the sizes of individual specimens (or their colonies). An extract is given in Table 2.6.

Table 2.6 Extract from the abundance scales proposed by the Marine Nature Conservation Review for both littoral and sublittoral taxa from 1990 onwards.

	Size of individuals (or colonies)				
% cover	< 1 cm	1–3 cm	3–15 cm	> 15 cm	Density per m²
≥ 80	S				≥ 10,000
40–79	A	S			1000–9999
20–39	C	A	S		100–999
10–19	F	C	A	S	10–99
5–9	O	F	C	A	1–9
1–5	R	O	F	C	0.1–0.9
< 1 and density 0.01–0.09 per m²	L	R	O	F	0.01–0.09
< 1 and density 0.001–0.009 per m²		L	R	O	0.001–0.009
< 1 and density 0.0001–0.0009 per m²			L	R	0.0001–0.0009
< 1 and density < 0.0001 per m²				L	< 0.0001

Key: S = Superabundant; A = Abundant; C = Common; F = Frequent; O = Occasional; R = Rare; L = Less than rare

2.8 *Variation between and within quadrats

In Example 2.3, separate estimates (in that case, they were estimates of cover) were obtained from a number of subregions (in that case, transects) within a quadrat. Because the quadrat was not completely homogeneous, the separate estimates were not identical. Suppose now that there are Q quadrats, with each quadrat containing n distinct subregions. Assuming that the variability between subregions is the same for all quadrats, the pooled estimate of this common *within-quadrat variance*, σ_w^2, is

$$s_w^2 = \frac{1}{Q(n-1)} \sum_{j=1}^{Q} \left(\sum_{i=1}^{n} c_{ji}^2 - \frac{1}{n} c_{j0}^2 \right),\tag{2.12}$$

where the $c_{j1}, c_{j2}, ..., c_{jn}$ are the n separate estimates of cover in quadrat j, and

$$c_{j0} = \sum_{i=1}^{n} c_{ji},$$

is the sum of the subregion values.

The sample variance of the quadrat totals $c_{10}, c_{20}, ..., c_{Q0}$ is

$$s_q^2 = \frac{1}{(Q-1)} \left(\sum_{j=1}^{Q} c_{j0}^2 - \frac{1}{Q} c_{00}^2 \right),\tag{2.13}$$

where

$$c_{00} = \sum_{j=1}^{Q} c_{j0},$$

is the overall total of the nQ values. Just as the subregions may vary within a quadrat, quadrats may vary within the entire region being sampled. Denoting the associated *between-quadrat variance* by σ_b^2, it can be shown that s_q^2 is an estimate of $n^2\sigma_b^2 + n\sigma_w^2$, so that an estimate of σ_b^2 is provided by

$$s_b^2 = \frac{1}{n^2} \left(s_q^2 - ns_w^2 \right).\tag{2.14}$$

The ratio s_q^2/ns_w^2 provides an indication of whether there is substantial quadrat-to-quadrat variation beyond that to be expected as a consequence of the variation in the subregions.[6] The overall estimate of cover is

$$\bar{c} = \frac{1}{nQ} \sum_{j=1}^{Q} \sum_{i=1}^{n} c_{ji},$$

with variance estimated by

$$\frac{1}{n^2 Q} s_q^2,$$

so that a very approximate 95% confidence interval for cover in the region is

$$\left(\bar{c} - 2\sqrt{\frac{1}{n^2 Q} s_q^2}, \ \bar{c} + 2\sqrt{\frac{1}{n^2 Q} s_q^2} \right).\tag{2.15}$$

Example 2.8: Alaskan shrubs (cont.)

Figure 2.18 illustrated the locations of shrubs in the five transect subregions of a quadrat. The results[7] for that quadrat are reported in the following table along with those for three other randomly chosen quadrats:

Quadrat, j	Separate cover estimates, $\{c_{ji}\}$					Total, c_{j0}	c_{j0}^2
1	0.1075	0.0817	0.0471	0.0555	0.0297	0.3215	0.10336
2	0.0810	0.0939	0.0929	0.1041	0.1116	0.4835	0.23377
3	0.0344	0.0094	0.0085	0.0373	0.0166	0.1062	0.01128
4	0.0544	0.0904	0.0491	0.1029	0.0561	0.3529	0.12454
Total	$\sum c_{ji}^2 = 0.10200$					1.2641	0.47295

Then

$$s_w^2 = \frac{1}{4(5-1)}\left(0.10200 - \frac{1}{5} \times 0.47295\right) = 0.000463,$$

and

$$s_q^2 = \frac{1}{3}\left(0.47295 - \frac{1}{4} \times 1.2641^2\right) = 0.024488.$$

Thus

$$s_b^2 = \frac{1}{25}(0.024488 - 5 \times 0.000463) = 0.000887,$$

and the variance between quadrats is seen to be about twice the variance within a quadrat.

The overall estimate of cover is $\bar{c} = 1.2641/20 \approx 6.3\%$, with an approximate 95% confidence interval of (3.2%, 9.5%).

3. Points and lines

A point can be regarded as an extremely small quadrat that provides information on frequency ('Is species X present at this spot?') and cover type ('What is present at this point?'). Much of the discussion concerning quadrats is therefore relevant, though the context must be considered. For example, when using satellite photographs, randomly positioned points, possibly using strata (see Section 2.3), may be entirely appropriate. However, in the field, using random points may be impractical and a more organized approach may be required. For example, in the context of underwater sampling, a line, with knots at regular intervals, may be randomly located on the underwater surface. Sampling takes place at the points indicated by the knots. This is called *point-intercept sampling*. To avoid bias, whichever sampling procedure is used, the points should be as small as is practicable (Wilson, 1963).

Point sampling has also been used to study association between species (Yarranton, 1966), and as the basis for identifying the type of spatial pattern (random, regular, or clustered) that a species exhibits (see e.g. Upton and Fingleton, 1985).

3.1   The point quadrat frame

The point quadrat frame (also known as a *Levy bridge*) is a small portable frame standing on two legs. Between the legs, is either a line of pins, or a line of holes through which a pin may be inserted. Typically, there are ten pin locations, at intervals of about two inches. The first detailed description of the use of the frame is by Levy and Madden (1933). If a ten-pin frame is placed at ten random locations in a study area, then the total number of occasions on which a plant is hit by a pin, gives the percentage cover for the region.

The ten-pin frame has the advantage that, when the frame is in position, ten observations are quickly obtained. However, because spatial autocorrelation is usually strongly positive, the ten observations will be much more similar to one another than ten widely separated observations would be. Goodall (1952) and Kemp and Kemp (1956) demonstrated that, for this reason, more precise results would be obtained by using the frame with fewer pins (but at a larger number of locations). Rothery (1974) took account of the potential costs of moving the frame to more positions and concluded that it was indeed better to use just one or two pins per location.

3.2   Line-intercept sampling (LIS)

This is a method of sampling the particles present in a region of interest, by observing those particles that overlap a line randomly placed in the region. The method, which is also called *line-intersect sampling* (since it involves the intersection of a one-dimensional line with the two-dimensional 'footprint' of a particle), was introduced to ecologists by Canfield (1941). The 'particles', which must be essentially stationary, may be terrestrial (e.g. plants, bare ground, trees, logs, debris, animal signs, animal dens), above ground (e.g. bird nests, tree canopy), or below the surface of water (e.g. coral). If the measurements concern ground cover, then a tape is stretched taut on the ground. If tree cover is being assessed, then the tape will be stretched taut above the ground cover. In the context of underwater sampling with an appreciable current, having chosen a starting point at random from a feasible region, it may be necessary to lay the transect parallel to the current (Beenaerts and Vanden Berghe, 2005).

LIS has many potential purposes. It can be used to estimate the number of particles present, the proportion of the region occupied by the particles (the *coverage*), the average age, weight, or other such quantities. If the 'particles' are actually gaps in planted crops, then the quantity estimated may be called the *stand loss*. In a row of crops the interval in which plants are missing might be termed a *skip* (Willers et al., 1992), while the interval between plants in a natural habitat might be called a *fetch* (Kuehl, McClaran, and van Zee, 2001).

---

❶ *Advice on data collection*

Suppose that a study area containing the N particles, $P_1, P_2, ..., P_N$, is sampled with n transects which have lengths $L_1, L_2, ..., L_n$. The random locations and orientations of the transects may be determined as follows. First, use random numbers to select a location within the study area as one end of the transect. Next select a random angle in the range $(0°, 360°)$ to determine the orientation of the transect. Since it is not necessary for the transects to be of equal length, if a transect would otherwise extend outside the study region, then it can be shortened appropriately.

Various rules have been proposed to deal with cases such as a particle overlapping the end of a transect, or a particle overlapping the edge of the study area, or both (see e.g. McDonald, 1980). With particles that are small relative to the study area, these complications may reasonably be ignored.

Suppose that, collectively, the n transects encounter m of the N particles. Consider the quantity $\hat{\gamma}$ given by

$$\hat{\gamma} = \sum_{j=1}^{m} x_j \left/ \sum_{i=1}^{n} L_i \right. . \tag{3.1}$$

In this expression the numerator is simply the total length of the intersections of particles with the transect lines, while the denominator is the total transect length. Thus $\hat{\gamma}$ is the overall proportion of the transect lengths that is intersected by particles. Since the transects are collectively representative of the entire study region, $\hat{\gamma}$ is intuitively an estimate of the proportion of the region intersected (i.e. covered) by particles. Lucas and Seber (1977) showed that $\hat{\gamma}$ is indeed an unbiased estimate of the coverage, γ.

Lucas and Seber (1977) also showed that an unbiased estimate of the population density, ρ, is given by

$$\hat{\rho} = \sum_{j=1}^{m} \frac{1}{w_j} \left/ \sum_{i=1}^{n} L_i \right. , \tag{3.2}$$

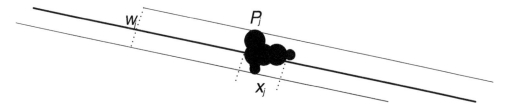

Figure 3.1 Line-intercept sampling. The irregularly shaped object, P_j overlaps the transect line. The quantity w_j is the width of P_j (measured perpendicular to the transect line). The quantity x_j is the length of the intersection of P_j with the line.

where w_j is measured as the perpendicular distance between tangents to particle P_j that are parallel to the transect line (see Figure 3.1). A transect has a greater chance of intercepting a wide particle than a narrow particle; the factor $1/w_j$ corrects this imbalance for particle P_j.

Let x_j be some other attribute of particle P_j that is of interest (for example, volume, weight, or age). McDonald (1980) observed that

$$\hat{\alpha} = \sum_{j=1}^{m} \frac{x_j}{w_j} \bigg/ \sum_{i=1}^{n} L_i , \tag{3.3}$$

is an unbiased estimate of $\sum_{j=1}^{N} x_j/A$, where A is the area of the region being sampled. Taking account of the orientations of the intersected particles, the average value of x is therefore estimated by

$$\bar{\alpha} = \sum_{j=1}^{m} \frac{x_j}{w_j} \bigg/ \sum_{j=1}^{m} \frac{1}{w_j} . \tag{3.4}$$

As an estimate of the average particle size, McDonald (1980) suggested using

$$\bar{\mu} = \sum_{j=1}^{m} y_j \bigg/ \sum_{j=1}^{m} \frac{1}{w_j} . \tag{3.5}$$

For the case of long thin particles, such as logs or downed trees, de Vries (1973) showed that $\hat{\rho}$ may be calculated using

$$\hat{\rho} = \pi \sum_{j=1}^{m} \frac{1}{l_j} \bigg/ 2 \sum_{i=1}^{n} L_i , \tag{3.6}$$

where l_j is the length of particle (log) j. For an attribute x, the corresponding result (de Vries, 1973) is

$$\hat{\rho} = \pi \sum_{j=1}^{m} \frac{x_j}{l_j} \bigg/ 2 \sum_{i=1}^{n} L_i . \tag{3.7}$$

Kaiser (1983) showed that all the previous results could be regarded as special cases of a general formula. If the use of a set of independent random transects is not feasible, then these formulae will continue to apply if the particles are randomly placed and oriented.

ℹ️ *Advice on data collection*

'Most plant species have some gaps in their canopies, such as bunch grasses with dead centers or shrubs with large spaces between branches. Because observers treat gaps differently, rules for dealing with gaps must be clearly defined. One solution is for the observer to assume a plant has a closed canopy unless a gap is greater than some predetermined width. We recommend that gaps less than 2 inches (5 cm) be considered part of the canopy.' USDA Forest Service Gen. Tech. Rep. RMRS-GTR-164-CD. 2006 (J. F. Caratti)

While a series of equi-spaced parallel transects is an efficient way of ensuring that a study area is covered, it could give biased results if most particles have the same orientation. Similar orientations would be expected for fallen trees after a hurricane, and underwater particles affected by the water current. In such situations a solution might be to choose a random direction, and then use two series of transects, one parallel to the selected direction, and one at right angles to that direction. Gregoire and Valentine (2003) discuss the use of L-shaped transects, where it is necessary to follow careful rules to distinguish between complete and partial intersections. Affleck, Gregoire, and Valentine (2005) extend the discussion to multisectioned transects including triangular transects which appeal because the researcher following the transect path will return to the starting point.

In a study of the efficiency of different transect configurations for estimating the amount of coarse woody debris, Nemec and Davis (2002) concluded that, in general, a few long transects gave more reliable results than many short transects.

In the underwater context, Leujak and Ormond (2007) noted that, when using line-intercept sampling, there may be inflated estimates of cover if the physical line follows the three-dimensional contours of 'massive and branching hard corals', rather than being truly linear in three-dimensional space.

3.2.1 Chain intercept transect sampling

This variant of line-intercept sampling was described by Dollar (1982) in the context of measuring the abundance of corals. Along the transect line, a chain is placed on the surface of the coral community. The number of links touching each piece of coral is recorded. Comparison of the length of chain with the surface length provides a measure of the *rugosity* (the surface roughness) of the coral.

3.3 Point-count transect sampling

When sampling deep under water, only a limited time is available to the sampling diver. This may make line-intercept sampling impractical. As an alternative, in the context of coral reefs, Roberts et al. (2016) suggested using sampling points at regular intervals along a randomly placed transect. Each sample would consist of recording the species of a specified number of corals adjacent to the sampling point. Their pragmatic suggestion was to work outwards from the sampling point in a spiral, with each new coral being chosen as that apparently closest to its predecessor in the chosen direction.

Table 3.1 Extracts from the results reported by Roberts et al. (2016) in a comparison of the use of the line-intersect and point-count methods.

	Line-intercept				Point-count			
No. of transects	1	2	3	4	1	2	3	4
No. of specimens	11	21	30	48	12	24	36	48
No. of species	6	6	8	14	11	19	24	33

The distance between successive sampling points would be chosen to avoid encountering any specimen more than once.

The method is somewhat subjective, with potentially different corals being chosen dependent on the direction of the spiral. Since no size measurements are made, the method reports only the relative frequencies with which species are found, not their overall contributions to the coral reef. The method was found to be much superior to line-intercept sampling in its ability to detect different species. Roberts et al. (2016) attribute this to the fact that 'the transect line is almost always unable to follow the reef contours precisely, missing most of the complex habitat'. Table 3.1 illustrates the difference: after 48 specimens have been recorded by each method, the point-count method has encountered more than twice the number of varieties encountered using line-intersect transects.

3.3.1 The wheel-point method

This method, first described by Tidmarsh and Havenga (1955), was originally used in the grassland of South Africa. The method uses a pair of rimless wheels on the end of a long handle. The wheels are rolled over the ground on their spokes; the sampling points are the locations where a particular sharpened spoke touches the ground (or plant). The method ensures that sampled points will be well separated (at distances equal to the circumference of the wheel). As described, the method required three workers, one to push the wheels, one to identify any vegetation struck by the sharpened spoke, and one to record results. The authors suggested using about 2000 sampling points.

3.3.2 The step-point method

This simple method for measuring cover, introduced by Evans and Love (1957), is widely used in North American rangelands where there are few obstacles. An observer samples a region by walking along a series of paths, taking observations at regular intervals (typically every 5 or 10 paces). If all that is required is a general idea of the condition of the land surveyed, then it will suffice to use one person with a clearly marked boot. At each observation point, a note is made of the terrain type (e.g. grass, bare earth, rock) adjacent to the boot mark. To avoid bias, it is preferable to predetermine the path locations using, for example, a series of parallel paths at fixed distances from one another. For reliable results, several hundred measurements are required, but the simplicity of the method makes this a practical requirement. Cover is then calculated as the proportion of the measurements satisfying the requirements.

For more sophisticated measurements, that record plant type, Evans and Love (1957) suggested using a boot with a notch cut in its toe. They proposed inserting a thin vertical rod into the notch, and recording the first plant touched, starting from the top of the rod (for a study of foliar cover or canopy cover) or from the bottom of the rod (for a study of ground cover). However, Wilson (1960) observed that using a vertical rod would 'underestimate erect-leaved species and exaggerate the contribution of species having more nearly horizontal leaves'. His theoretical study suggested that inclining the rod at an angle of about 30° to the ground gave more reliable estimates.

ⓘ Advice on data collection

At an observation point, if the boot is not placed flat on the ground, there is less chance of plant disturbance biasing results.

In sparsely vegetated areas, occasions on which the rod touches any plant will be few and far between. Evans and Love (1957) suggested that, in this case, if the interest is in the extent to which the various species are present, then the plant nearest in a forward direction should be recorded. However, Strauss and Neal (1963) demonstrated that this results in large plants being over-recorded. They suggested that it was preferable to record the plant, in a forward direction, whose centre was closest to the sampling point.

An approximate 95% confidence interval for the amount of cover, can be calculated using Equation (2.6), with n denoting the number of sample points, and r the number of points with cover present. A corresponding calculation would be used to calculate confidence intervals for features such as the incidence of rock or bare ground. However, if the question of interest is the proportion of cover supplied by a particular species, then n is the number of sample points with cover of any type, and r is the number of points where the species of interest is present.

Example 3.1: Californian grassland

Evans and Love (1957) presented the results for nine observer pairs who each sampled the same one-acre plot. The observers had no previous experience of the step-point method. An extract from their results is presented in Table 3.2.

Considering only the data obtained by the first observer pair ($n = 100$ and $r = 43$), and using Equation (2.6), cover is estimated by $\tilde{p} = (43 + 2)/(100 + 4) \approx 0.43$ and the approximate 95% confidence interval is found to be (0.34, 0.54). Although the results

Table 3.2 Results from the 1957 study by Evans and Love of the efficacy of the step-point method. Each observer pair collected data at 100 sampling points in the same one-acre field.

	Nine observer pairs								
Points with cover of any type	43	31	41	33	43	34	31	43	39
Points with *Erodium botrys* present	6	4	6	3	5	3	3	6	5
Erodium botrys as % of total cover	14	13	15	9	12	9	10	14	13

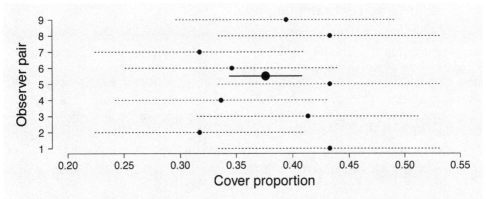

Figure 3.2 The values of $\tilde{p}$ and the approximate 95% confidence bounds for the cover proportion for each of nine observer pairs, together with the result that would be obtained if the nine sets of observations had been obtained from separate regions.

for the nine observer pairs vary, their approximate confidence intervals overlap as can be seen in Figure 3.2. The figure also shows the much narrower interval that would result if nine separate regions had been sampled and the separate results had been aggregated (so that $n = 900$ and $r = 338$).

The same procedure can be used when considering the proportion of the overall cover that is due to *Erodium botrys*.

For the first observer pair, $n = 43$ and $r = 6$, so that $\tilde{p} = (6 + 2)/(43 + 4) \approx 0.17$. This is appreciably greater than the observed proportion ($p = 6/43 \approx 0.14$).[1] Although the difference is large for each pair, the difference is slight when the data are aggregated (which would be the sensible procedure if the observers had sampled different regions), since then $p = 41/338 \approx 0.121$ and $\tilde{p} = 43/342 \approx 0.126$. Thus $\tilde{p}$ for the combined data is smaller than all the separate $\tilde{p}$ values. The very wide approximate 95% confidence intervals, illustrated in Figure 3.3, are also a consequence of the small value for n and indicate that, to obtain reasonably precise estimates, it is necessary to obtain several hundred observations.

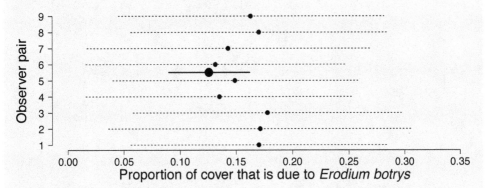

Figure 3.3 The values of $\tilde{p}$ and the approximate 95% confidence bounds for the proportion of cover accounted for by *Erodium botrys*.

3.3.3 Point-intercept sampling

One criticism of the step-point method (Section 3.3.2) is that both the distances between sample points, and the locations chosen for the sample points, may be influenced by the terrain, leading to bias in the results. The point-intercept method avoids such biases, by using a tape (or equivalent) stretched between two points. Before the tape is positioned, the sampling points are clearly marked on the tape. Since the sample path must be traversed three times (to erect the tape, sample, and remove the tape), in the field this is a slower procedure than the step-point method. However, it is faster than line-intercept sampling (Section 3.2), and is commonly used in underwater sampling where the underwater time may be limited.

4. Distance methods

Instead of using well-defined sample plots, the methods introduced in this chapter use combinations of distances between sample points and nearby plants,[1] and between those plants and their neighbours. The methods used in this chapter should not be confused with the distance methods introduced in Chapter 8.

All the methods work well with plants that are distributed at random in the study area, but many are not reliably accurate for non-random arrangements. When the purpose of sampling is not simply to assess the number of individuals present, but also, for example, to assess the quality and quantity of timber stocks, then distance sampling is attractive since, by definition, at least one plant will be observed from each sampling point. By contrast, with a sparse pattern, a randomly placed fixed-size quadrat might fail to include a single plant.

The first accounts of the use of inter-plant distances date to the late 1940s and the following decade. Influential papers are those of Clark and Evans (1954), who popularized the idea of using distances between neighbouring plants to provide information about spatial pattern, and Cottam and Curtis (1956), whose focus was on determining plant density.

Some of the formulae presented in this chapter display slight differences from those in the literature. There are two reasons for this. The first is that the formulae might be described as 'fiddly' and as a consequence typographical errors are easily made. The second reason is that the adjustment needed for the plant-to-nearest-neighbour distances, correctly used by Cottam and Curtis (1956), has often not been carried through to the related estimators subsequently proposed.

4.1 Spatial patterns

Whereas the effectiveness of quadrat methods is unaffected by the type of plant pattern, the behaviour of distance methods is critically dependent on the pattern type. To illustrate this dependence, as the alternative distance methods are introduced, they will be assessed using the results from simulated data of four types: random, gradient, regular, and cluster.[2]

Examples of each data type are shown in Figure 4.1.

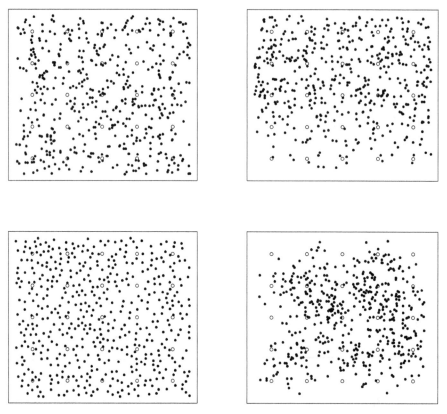

Figure 4.1 Examples of the four types of pattern used in the simulations. Clockwise from top left, these are random, gradient, clustered, and regular. Each diagram shows 500 plants (●), and the grid of 25 sampling points (○).

4.2   Locations for sampling points

The basic criteria for the choice of sampling points is that:

1.   The sampling points should be chosen independently of the locations of the plants being studied.
2.   They should thoroughly represent the region of interest.

These criteria are often interpreted as implying that the sampling points should be chosen at random. This is easily achieved in computer studies, but could lead to difficult routing in the field. However, there is a more serious difficulty inherent with truly random points: the result will be that some regions will include many sampling points, while others, of the same size, will have few sampling points. The second criterion will not have been satisfied.

One solution is to lay down a regular grid of sampling points, so that:

•   each point is effectively sampling the same proportion of the study area;
•   each point is equidistant from its nearest neighbouring sampling point;
•   no point is too close to the edge.

This is a practical arrangement that simplifies the sampling logistics. It also usefully provides information about the type of spatial pattern, since each sampling point may provide its own estimate of population density, as well as contributing to a pooled estimate. When the area sampled is inhomogeneous, the average of the individual estimates may be more reliable than the pooled estimate, while the variability of the individual estimates provides a useful measure of the precision of the averaged estimate.

4.2.1 Edge effects

The methods discussed in this chapter require the determination of one or more distances. Either the distance from a sampling point to a nearby plant, or from a plant to its nearest neighbour (or both). Generally, although there may be practical difficulties in establishing which of a number of alternative plants is the nearest, and in accurately determining that distance, it is at least clear what distance should be measured. The exception occurs when an edge of the study region is closer than the nearest visible plant. If the distance to the nearest plant in the region is used, then this may be greater than the distance to a plant outside the region, so that the density estimate is biased downwards. If no value from this point is reported, then that would imply that measurements near the edge were made only where plants were relatively common, so the density estimate would be biased upwards.

When using the methods of this chapter (except for those discussed in Section 4.9) it is assumed that the grid of sampling points is positioned so that edge effects can be neglected. Other methods for dealing with edge effects were discussed in Section 2.4.

4.3 Simple point-to-plant measures

Suppose a sampling point is placed in a study region that contains randomly placed plants at an average intensity of ρ per unit area. Define X as the distance from the point to its nearest neighbouring plant. If X is equal or greater than some value x, then this implies that there are no individuals in a region of area πx^2, for which the mean number of plants would be expected to be $\rho\pi x^2$ (see Figure 4.2).

Figure 4.2 A sampling point (○) is placed in a study area containing several plants (●). The nearest plant is at a distance x. There are no plants in the shaded region which has area πx^2.

Since the plants are distributed at random, the probability of there being no plants in a circle of radius x, $P(X > x)$, is the probability of an observation of zero from a Poisson distribution with mean $\rho\pi x^2$. This probability is $\exp(-\rho\pi x^2)$. The corresponding probability density function is

$$f(x) = 2\rho\pi x \exp(-\rho\pi x^2),$$

and the expected values of X and X^2 are respectively

$$E(X) = 1/(2\sqrt{\rho}), \tag{4.1}$$

and

$$E(X^2) = 1/(\rho\pi). \tag{4.2}$$

Using n sampling points, and denoting the distance from sampling point i to its nearest plant by x_i ($i = 1, 2, \ldots, n$), $E(X)$ is estimated by $\sum_{i=1}^{n} x_i/n$. Substitution in Equation (4.1), and rearrangement, leads to the estimate of ρ suggested by Clark and Evans (1954):

$$\widehat{\rho}_x = n^2 \left/ 4 \left(\sum_{i=1}^{n} x_i \right)^2 \right. . \tag{4.3}$$

An alternative is to use $\sum_{i=1}^{n} x_i^2/n$ as an estimate of $E(X^2)$. This suggests estimating ρ by $n/\left(\pi \sum_{i=1}^{n} x_i^2\right)$. Moore (1954) showed that the bias in this estimator is easily corrected by multiplying it by $(n-1)/n$ to give:

$$\widehat{\rho}_{x2} = (n-1) \left/ \left(\pi \sum_{i=1}^{n} x_i^2 \right) \right. , \tag{4.4}$$

with (for a random plant pattern) variance equal to $(\widehat{\rho}_{x2})^2 /(n-2)$.

The previous results assume that the plants are placed at random. A more usual case is that plants occur in clusters, so that there are large areas of lower than average density, and small areas of higher than average density. Any arrangement of sampling points will therefore usually result in the majority of sampling points falling in the regions of low density. With clustered plants the result is therefore likely to be an under-estimate of plant density.

4.4   Using the distance to the kth nearest plant

Equation (4.4) is a special case of a general class of estimators suggested by Morisita (1957):

$$\widehat{\rho}_{(k)} = (kn-1) \left/ \left(\pi \sum_{i=1}^{n} x_{(k)i}^2 \right) \right. , \tag{4.5}$$

where $x_{(k)i}$ is the distance from sampling point i to its kth nearest plant (Figure 4.3). Thus x_i is now written as $x_{(1)i}$.

For a random plant pattern, the variance of this estimator is

$$\left(\widehat{\rho}_{(k)}\right)^2 /(kn-2) .$$

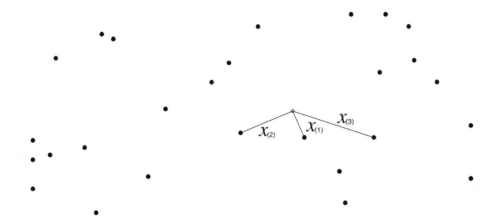

Figure 4.3 Some methods for estimating plant density use $x_{(2)}$, the distance to the second nearest neighbour, $x_{(3)}$, the distance to the third nearest neighbour, and so on.

Pollard (1971) noted that the k in the denominator implies that the variance of the estimate is approximately halved by using the second nearest tree rather than the nearest, with further reductions gained by using larger values of k.

Using $x_{(k)}$ is often referred to as *k-tree sampling*. It is particularly convenient when the experimenter needs to record several characteristics of the 'plants' under investigation (e.g. species, size, etc.) and not simply their location. If fixed-radius circular quadrats were used instead, then the radius might be too large in a clumped region or too small in a sparse region. In either case the experimenter might feel that resources were being wasted.

To investigate the accuracy of the various distance methods, 1000 patterns of each of the four types were generated within a square study region, with each containing 500 plants. A regular grid of 25 sampling points was used (as illustrated in Figure 4.1).

Figure 4.4 summarizes the results using box-whisker plots (Section 1.2.3). Here a logarithmic scale is used, so that 0.5 corresponds to an estimate of 250, 1 corresponds to an estimate of 500, 2 to an estimate of 1000, etc. The ideal estimate would therefore be represented by a narrow box with centre 1 and with short whiskers.

For random patterns, all seven estimators are unbiased (i.e. they are centred on 1), but with variability that decreases steadily as k increases (ever narrower boxes and shorter whiskers). Prodan (1968) advocated the use of the distance to the sixth nearest plant. The results summarized in Figure 4.4 support this suggestion, since the boxplot for $k = 6$ is close to ideal. If the choice of $k = 6$ is impractically large, then the largest practical value should be used.

For regular patterns, both $\hat{\rho}_x$ and $\hat{\rho}_{x2}$ generally overestimate the number of plants, but, as k increases, so the extent of overestimation by $\hat{\rho}_{(k)}$ reduces. For patterns with a gradient, and for clustered patterns, the reverse occurs, with considerable under-estimation by $\hat{\rho}_x$ and by $\hat{\rho}_{(k)}$ for low values of k.

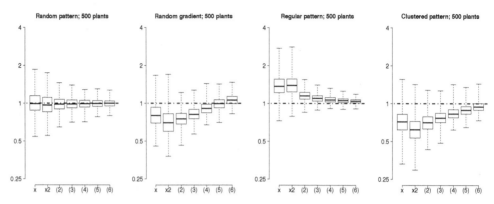

Figure 4.4 Box-whisker plots comparing the accuracy of alternative estimators of plant density using the distance from a sampling point to a nearby plant. '(k)' signifes the use of the kth nearest neighbour.

4.4.1 Non-random patterns

Morisita (1957) addressed the problem of patterns with varying density. He observed[3] that 'the distribution of biological individuals … tends to be aggregatedly distributed' rather than random. He suggested that the study area 'can be divided theoretically into several small fractions in which no aggregated distribution can be observed' (ideally with the plants being randomly distributed within each fraction). He noted that 'the density is not necessarily the same in the different fractions' and therefore proposed estimating the density separately for each fraction using a single observation in each case.

Substituting 1 for n in Equation (4.5) gives the density estimate for the single fraction surrounding sampling point P_i as

$$\widehat{\rho}_{P_i} = \frac{k-1}{\pi x_{(k)i}^2}.$$

Averaging over the n separate fractions gives the estimate for the entire region as

$$\widehat{\rho}_{(k)a} = \frac{1}{n}\sum_{i=1}^{n}\widehat{\rho}_{P_i}. \tag{4.6}$$

An indication of the overall applicability of the estimate is provided by examining the variability of the n individual estimates. Their variance is estimated by

$$s_\rho^2 = \frac{1}{n-1}\left\{\sum_{i=1}^{n}\left(\widehat{\rho}_{P_i}\right)^2 - n\left(\widehat{\rho}_{(k)a}\right)^2\right\}, \tag{4.7}$$

so that an approximate 95% confidence interval (which reflects both sampling uncertainty and the extent of the variations in plant density across the area studied) is provided by

$$\left(\widehat{\rho}_{(k)a} - 2s_\rho/\sqrt{n},\ \widehat{\rho}_{(k)a} + 2s_\rho/\sqrt{n}\right). \tag{4.8}$$

The box-whisker plots summarizing the computer results using $\widehat{\rho}_{(k)a}$ are illustrated in Figure 4.5. Comparison with Figure 4.4 shows that, for patterns with varying local density (the gradient and clustered patterns), the median value for the $\widehat{\rho}_{(k)a}$ estimators

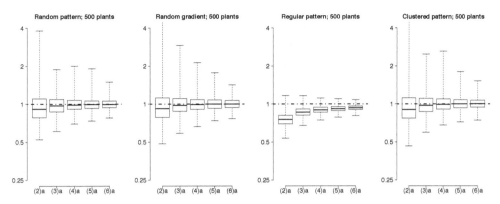

Figure 4.5 Box-whisker plots comparing the accuracy of alternative estimators of plant density using the average distance from a sampling point to a nearby plant (ignoring edge effects).

may be, on average, closer to the target value, However, this advantage is offset by the fact that this group gives more variable results than the $\hat{\rho}_{(k)}$ estimators. Whichever of $\hat{\rho}_{(k)}$ and $\hat{\rho}_{(k)a}$ is chosen, it is evident that k should be chosen to be as large as is practicable. A bonus when using $\hat{\rho}_{(k)a}$ is that the precision of the estimate and the variability in the plant density across the region are made explicit using Equation 4.8.

Example 4.1: Mangroves in Kenya

According to Hijbeek et al. (2013) 'progress [in mangrove forests] is particularly difficult when trying to get deeper into the forest as often climbing over tree roots is required. Laying plots in a mangrove forest can range from being extremely time consuming up to unfeasible.' For this reason, Cintrón and Schaeffer-Novelli (1984) had suggested that a plotless method should be used in preference to quadrats.

Hijbeek et al. (2013) compared several distance estimators using four sets of mangrove data (illustrated in Figure 4.6)[4] and various types of simulated data. Sites 1 and 2, which include several species of mangrove, show a marked variation in plant density along their lengths. Sites 3 and 4, which were populated by a single mangrove species (*Avicennia marina*, the white mangrove), differ by their canopy type (closed in Site 3; open in Site 4).

Hijbeek et al. (2013) found that both $\hat{\rho}_x$ and $\hat{\rho}_{x2}$ substantially under-estimated the number of mangroves present. For the artificial data sets, for which they extended their investigation to include $\hat{\rho}_{(k)}$ and $\hat{\rho}_{(k,4)}$ with $k = 2$ or 3, they again found no satisfactory estimator.

The results in Table 4.1 confirm that the $\hat{\rho}_{(k)}$ series of estimators do provide extreme under-estimates for the heterogeneous Sites 1 and 2.[5] The estimates using the $\hat{\rho}_{(k)a}$ series are reasonably accurate.

For Site 1, Figure 4.7 compares the estimate provided by each sampling point with the actual number of mangroves in the subregion sampled by that point. There are gross errors in individual estimates, but there is no overall bias.

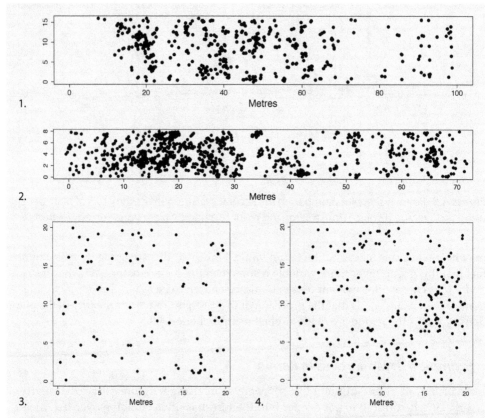

Figure 4.6 The locations of mangroves in four sites in a mangrove forest in Gazi Bay, Kenya. Sites 1 and 2 include several species with a marked variation in intensity across each. The mangroves in sites 3 and 4 are all *Avicennia marina* (the white mangrove) with Site 3 being a region with a closed canopy, whereas Site 4 has an open canopy.

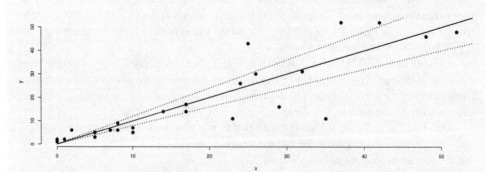

Figure 4.7 Comparison of the estimated counts using $\widehat{\rho}_{(6)a}$ with the true values in each section of mangrove Site 1. The line of equality is shown, together with dotted lines indicating 20% deviations from the true number.

Table 4.1 Estimates of abundance for the four mangrove sites illustrated in Figure 4.6. Point estimates within 20% of the true value are shown in bold. Those approximate 95% confidence intervals for the $\hat{p}_{(k)a}$ series that enclose the true value are given in bold.

| Site | Estimate | | | | | | No. of trees |
	$\hat{P}_{(3)}$	$\hat{P}_{(4)}$	$\hat{P}_{(6)}$	$\hat{P}_{(3)a}$	$\hat{P}_{(4)a}$	$\hat{P}_{(6)a}$	
1	103	124	149	350	**404**	**461**	472
2	450	474	474	1679	**1140**	**1157**	990
3	**75**	**79**	**74**	66	**76**	**81**	85
4	170	177	**183**	**193**	**192**	**198**	227

| Site | Approx. 95% confidence interval | | | No. of trees |
	$\hat{P}_{(3)a}$	$\hat{P}_{(4)a}$	$\hat{P}_{(6)a}$	
1	**(224, 475)**	**(259, 548)**	**(285, 638)**	472
2	**(210, 3149)**	**(610, 1670)**	**(655, 1659)**	990
3	**(50, 82)**	**(59, 92)**	**(58, 103)**	85
4	**(129, 257)**	**(139, 246)**	**(158, 239)**	227

4.5   The point-centred quarter method (PCQM)

With this method four measurements are obtained for each sampling point. These are the distances to the nearest plant in each quarter-plane (NE, NW, SE, SW) as illustrated in Figure 4.8.

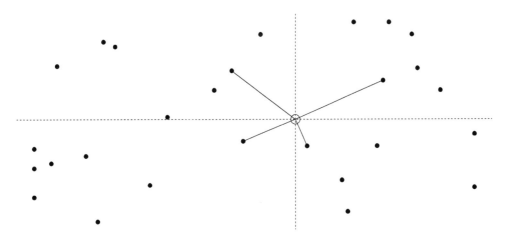

Figure 4.8 The point-centred quarter method. A sampling point (○) is randomly placed in a study area containing several plants (●). The distances to the nearest plant in each quarter are recorded.

An unbiased estimate of ρ was shown by Morisita (1957) to be given by

$$\widehat{\rho}_{pc} = 4(4n-1) \left/ \left(\pi \sum_{i=1}^{n} \sum_{j=1}^{4} x_{ij}^2 \right) \right. , \tag{4.9}$$

where $x_{i1}, \ldots, x_{i4}$ are the nearest-neighbour distances in the four quadrants surrounding sampling point i. For randomly distributed plants the estimate has variance $(\widehat{\rho}_{pc})^2/(4n-2)$.

In their simulations, Hijbeek et al. (2013) found that $\widehat{\rho}_{pc}$ seriously overestimated the density of regular patterns, but under-estimated the number of plants, when pattern density varied across the study region. Jost (1993) observed that $\widehat{\rho}_{pc}$ was unreliable when used with plants whose density varied across the study region. Apparently unaware that Morisita (1954) had also suggested its use, Jost proposed using instead the average-based estimator

$$\widehat{\rho}_{pc2} = \frac{12}{n\pi} \sum_{i=1}^{n} \left(1 \left/ \sum_{j=1}^{4} x_{ij}^2 \right. \right) . \tag{4.10}$$

With randomly distributed plants, Morisita (1954) proved the result, previously demonstrated empirically by Cottam, Curtis, and Hale (1953), that $\bar{x}$, the average of the four distances, is an unbiased estimate of $\sqrt{1/\rho}$. From this it follows that, with n sampling points, an estimate of ρ is provided by

$$1/\bar{x}^2 = 16n^2 \left/ \left(\sum_{i=1}^{n} \sum_{j=1}^{4} x_{ij} \right)^2 \right. .$$

This estimate is not used with complete data as it is known to be biased, but it underlies the method proposed by Warde and Petranka (1981) for dealing with the problem when there are empty quarters. With n_0 missing values, they suggested calculating the average of the available $(4n - n_0)$ distances and incorporating a scaling constant, C, with a value dependent on the proportion of missing values. Their table of values for C is closely approximated by the simple equation

$$C = 1 - 2.4p + 2.8p^2, \tag{4.11}$$

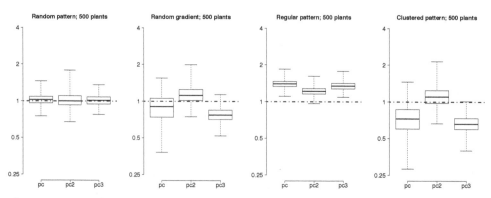

Figure 4.9 Box-whisker plots for the point-centred quarter method.

where $p = n_0/4n$. The resulting estimator is

$$\hat{\rho}_{pc3} = C/\bar{x}^2 = C(4n - n_0)^2 \bigg/ \left(\sum x\right)^2, \qquad (4.12)$$

where the summation is over the $(4n - n_0)$ measured distances.

Figure 4.9 suggests that the most successful of this group of estimators is $\hat{\rho}_{pc2}$.

Example 4.2: Mangroves in Kenya (cont.)

According to Hijbeek et al. (2013), the point-centred quarter method is the standard method for assessing numbers of mangroves. However, when they used $\hat{\rho}_{pc}$ with random sampling points for Sites 3 and 4 (they did not use it with Sites 1 and 2), they found it substantially under-estimated mangrove density. Table 4.2 gives estimates for all four mangrove sites, for all three point-centred quarter methods. The sampling points were those detailed previously. The results confirm the severe under-estimation by $\hat{\rho}_{pc}$, and show the expected superiority of the average-based $\hat{\rho}_{pc2}$. Note that the latter is arguably not as effective as $\hat{\rho}_{(4)a}$ which also involves measuring distances to four trees.

Table 4.2 Estimates of abundance obtained using the point-centred quarter method for the four mangrove sites illustrated in Figure 4.6. Point estimates within 20% of the true value are shown in bold.

	Site 1	Site 2	Site 3	Site 4
$\hat{\rho}_{pc}$	259	569	52	**187**
$\hat{\rho}_{pc2}$	315	**1022**	51	**227**
$\hat{\rho}_{pc3}$	153	564	45	159
True value	472	990	85	227

4.6 Angle-order estimators

The estimators $\hat{\rho}_{x2}$, $\hat{\rho}_{(k)}$, and $\hat{\rho}_{pc}$ are all special cases of a general class of estimators suggested by Morisita (1957). When the area around the sampling point is divided into q equal sections, with the distance being measured to the kth nearest plant in each section, then the resulting estimator is

$$\hat{\rho}_{(k,q)} = q(kqn - 1) \bigg/ \left(\pi \sum_{i=1}^{n} \sum_{j=1}^{q} x_{(k)ij}^2\right), \qquad (4.13)$$

where $x_{(k)ij}$ is the distance from the ith sampling point to the kth nearest plant in the jth section. The estimator has variance $(\hat{\rho}_{(k,q)})^2/(kqn - 2)$ for a random plant pattern. The case $k = q = 1$ corresponds to $\hat{\rho}_{x2}$, the case $q = 1$ corresponds to $\hat{\rho}_{(k)}$, and the case $k = 1$ and $q = 4$ corresponds to $\hat{\rho}_{pc}$.

Engeman et al. (1994) report excellent results when using $\widehat{\rho}_{(3,4)}$. However, White et al. (2008) remark that 'in practice much time is spent deciding which is the third closest individual and into which quadrant an individual lies'.

Khan et al. (2016) note that the formula for $\widehat{\rho}_{(k,q)}$ is often given incorrectly. The principal error, which is the replacement of $q(kqn-1)$ by $nq(kq-1)$ in Equation (4.5), is also noted in the excellent discussion of the point-centred quarter method by Mitchell (2015).

4.7 Nearest-neighbour distances

If it were possible to choose a plant at random, and measure the distance to its nearest neighbour, then the results of Section 4.3 would apply to those distances. However, a plant cannot properly be selected at random without having a list of all the plants – in which case there would be no need to estimate the plant density! Instead, as an approximation to the selection of a random plant, a sampling point is placed at random in the study region and the nearest plant is identified. The distance then recorded, y, is the distance between that plant and its nearest neighbour. Figure 4.10 illustrates the procedure.

The figure also illustrates why the results of Section 4.3 are not appropriate. As previously, it is known that the circular region of area πy^2 that surrounds the chosen plant contains no other plants. However, it is already known that the shaded area around the sampling point contained no other plants. This overlap, which will always occur, is at its greatest when $y \geq 2x$.

Cottam and Curtis (1956) examined a set of hypothetical random populations; they concluded that, by analogy with Equation (4.3), an approximately unbiased estimate of plant density was provided by scaling up by about 1.44 (= 1.2^2) to give:

$$\widehat{\rho}_y = 0.36n^2 \bigg/ \left(\sum_{i=1}^{n} y_i \right)^2, \tag{4.14}$$

where y_i is the nearest-neighbour distance corresponding to sampling point i. For a random pattern the variance of this estimator is given by $(\widehat{\rho}_y)^2/(n-2)$.

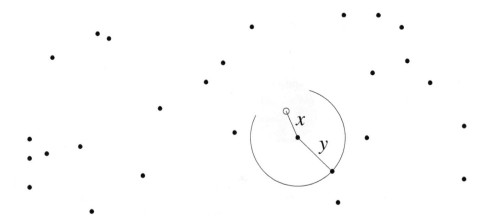

Figure 4.10 The nearest-neighbour method. A sampling point (○) is randomly placed in a study area containing several plants (●). The nearest plant is identified and the distance, y, from that plant to its nearest neighbour is determined.

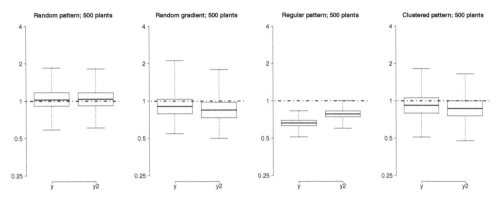

Figure 4.11 Box-whisker plots investigating the accuracy of the nearest-neighbour estimators.

In the same way, the equivalent of Equation (4.4) for a nearest-neighbour estimate is

$$\widehat{\rho}_{y2} = 1.44(n-1) \left/ \left(\pi \sum_{i=1}^{n} y_i^2 \right) \right.$$ (4.15)

Note that the version of this estimator that appears in the literature (often falsely attributed to Byth and Ripley (1973)) has $1.44(n-1)$ replaced by n.

In a regular pattern (a pattern where plants must keep their distance from one another), once a plant has been chosen, its neighbour is likely to be further away than would be expected with a random pattern of the same plant density. For this reason, estimators using only y-values are likely to seriously under-estimate the number of individuals in such a pattern. The extent of the under-estimation is illustrated in Figure 4.11.

4.8 Combined point-to-plant and nearest-neighbour measures

Diggle (1975) observed that, since x-based estimators under-estimate numbers in a clustered pattern, while y-based estimators often overestimate those numbers, an estimate that might be robust to deviations from randomness would be one that combined information from both x-values and y-values. He suggested combining Equations (4.4) and (4.15) to give:[6]

$$\widehat{\rho}_{xy} = 1.2(n-1) \left/ \left(\pi \sqrt{\sum_{i=1}^{n} x_i^2 \sum_{i=1}^{n} y_i^2} \right) \right. .$$ (4.16)

In a similar fashion Byth (1982) suggested combining Equations (4.3) and (4.14) to give:[7]

$$\widehat{\rho}_{xy2} = 0.3n^2 \left/ \left(\sum_{i=1}^{n} x_i \sum_{i=1}^{n} y_i \right) \right. .$$ (4.17)

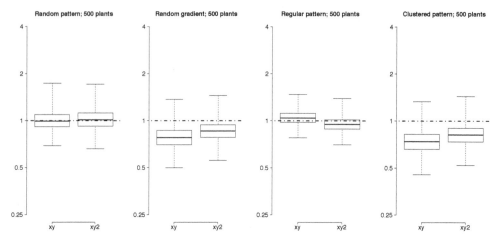

Figure 4.12 Box-whisker plots investigating the accuracy of measures combining point-to-plant and nearest-neighbour distances.

Figure 4.12 demonstrates that while these estimators are accurate for random or regular patterns, they under-estimate the average plant density when there are variations in density across the region of interest.

4.8.1 T-square estimators

As has been demonstrated, a difficulty with estimators that use y is caused by the variation in the overlap between the circle surrounding the sampling point, and the circle surrounding the nearest neighbour. Besag and Gleaves (1973) suggested avoiding this uncertainty by restricting attention to the half-plane away from the original sampling point, so that no overlap occurs. This is illustrated in Figure 4.13.

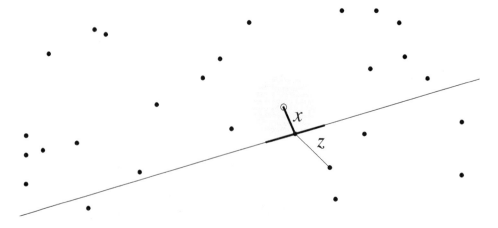

Figure 4.13 The T-square method. A sampling point (○) is randomly placed in a study area containing several plants (●). The nearest plant is identified. The distance, z, from that plant to its nearest neighbour in the half-plane away from the sampling point is determined.

Besag and Gleaves (1973) proposed using $\widehat{\rho}_z$ given by:

$$\widehat{\rho}_z = 2n \left/ \left(\pi \sum_{i=1}^{n} z_i^2 \right) \right. . \tag{4.18}$$

However, since estimates using z-values alone will, like those based on y-values, overestimate plant numbers in clustered patterns, Diggle (1975) suggested using estimators that combined information from both x and z:

$$\widehat{\rho}_{xz} = 2n \left/ \pi \left(\sum_{i=1}^{n} x_i^2 + \frac{1}{2} \sum_{i=1}^{n} z_i^2 \right) \right. , \tag{4.19}$$

$$\widehat{\rho}_{xz2} = n \left/ \pi \sqrt{\frac{1}{2} \left(\sum_{i=1}^{n} x_i^2 \right) \left(\sum_{i=1}^{n} z_i^2 \right)} \right. . \tag{4.20}$$

Byth (1982) examined many possible combinations of the various distances and concluded that the preferable combination was $\widehat{\rho}_{xz3}$ given by:

$$\widehat{\rho}_{xz3} = n^2 \left/ \left(2\sqrt{2} \sum_{i=1}^{n} x_i \sum_{i=1}^{n} z_i \right) \right. . \tag{4.21}$$

Silva et al. (2017) compared the T-square and point-centred quarter estimators with variant k-tree estimators in the context of forests in the Azores. They used the quadrat estimate as their 'benchmark'. They found that the T-square estimators outperformed the others with the best performing being $\widehat{\rho}_{xz3}$. Figure 4.14 confirms $\widehat{\rho}_{xz3}$ as being the most reliable, but suggests that it may seriously under-estimate plant density in clustered patterns.

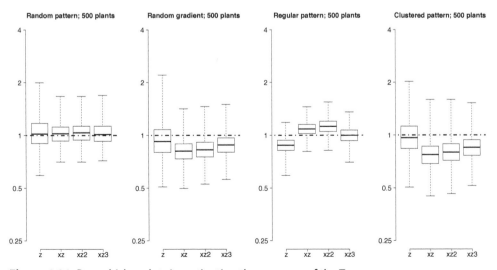

Figure 4.14 Box-whisker plots investigating the accuracy of the T-square measures.

4.9 Wandering methods

Catana (1953) observed that plants are more likely to be distributed in clumps rather than at random. He proposed a sampling procedure that, starting from a single sampling point, traverses the region of interest from plant to plant and thereby extracts a sequence of inter-plant distances. Catana proposed rules based on the mean and median of these distances to decide whether there was appreciable plant clumping. If clumps were diagnosed, then his procedure calculates average within-clump distances, average between-clump distances, clump sizes, and numbers of clumps, to arrive at an overall density estimate.

Catana's sampling procedure (illustrated in Figure 4.15) was as follows. From a randomly chosen sampling point on the edge of the study region, set a direction of interest (e.g. north) and identify the nearest plant in the quarter circle centred on this direction. Proceed to the identified plant and, treating the plant as a point, repeat the procedure. The result is now referred to as Catana's 'wandering-quarters' sampling procedure.

Diggle (1983) described the wandering-quarters method as 'an ingenious sampling procedure whose statistical potential appears not to have been tapped'. Hall, Melville, and Welsh (2001) subsequently generalized the idea to search segments of angle 2θ. They proposed the estimator

$$\widehat{\rho}_{w1} = (n - 1.5) \left/ \left(\theta \sum_{i=1}^{n} w_i^2 \right) \right. .$$ (4.22)

where w_i is a plant-to-plant distance and, for the wandering-quarters method, $\theta = \pi/4$. They outlined bootstrap-based corrections for bias based on prior information concerning the nature of the non-random pattern.

In the spirit of the wandering-quarters method, a 'wandering-forward' procedure is also presented, in which $\theta = \pi/2$. This procedure, which is illustrated in Figure 4.16, results in a shorter walk to obtain the same number of measurements. Simulations of random patterns suggest the following:

$$\widehat{\rho}_{w2} = (n - 1) \left/ \left(1.4 \sum_{i=1}^{n} w_i^2 \right) \right. .$$ (4.23)

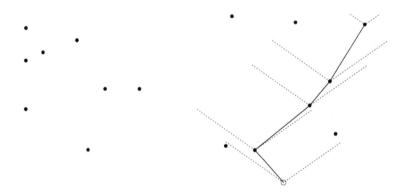

Figure 4.15 An example of Catana's wandering-quarters procedure ($\theta = \pi/4$).

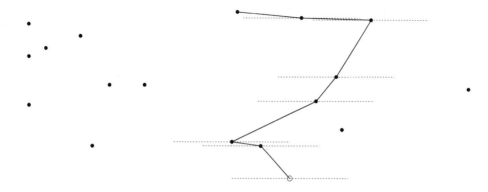

Figure 4.16 An example of the 'wandering-forward' procedure ($\theta = \pi/2$).

One practical advantage of the wandering procedures over those mentioned previously, is that the data collector is required to identify only one random location; furthermore, that location will be relatively easy to find as it will be on the perimeter of the region of interest. In the simulations reported here the wanderer was allowed to proceed from one side of the region of interest to the other. On average this resulted in 24 measurements for the wandering-quarters procedure (comparable to the 25 used with sampling points), but 43 measurements for the wandering-forward method. Figure 4.17 suggests that the procedures are reasonably unbiased for all pattern types. However, their variability is greater than for most of the alternatives considered. Unsurprisingly, since more measurements are taken, the wandering-forward procedure is the less variable.

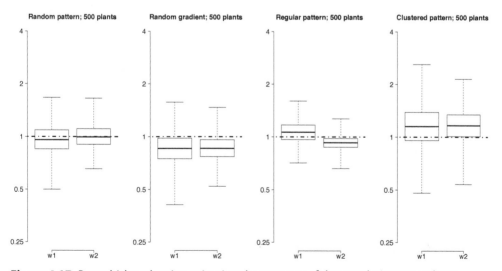

Figure 4.17 Box-whisker plots investigating the accuracy of the wandering procedures.

Example 4.3: Mangroves in Kenya (cont.)

The combination of variable plant density and the shape of the long narrow Sites 1 and 2 suggests that the wandering methods might struggle to obtain acceptable estimates. In Table 4.3 routes F and B indicate routes going forward from one end (the short edge in the case of Sites 1 and 2) or backwards from the other, in each case stopping after 25 trees have been encountered. The T routes traverse the sites from one end of the site to the other end, with no limit on the number of trees.

Superficially, the results are encouraging, with many of the estimates lying within 20% of the true value. However, for Site 1, the wandering-forward procedure is shown to be unacceptable, since, for the forward path, it gives rise to estimates that are three or four times the true value. For the backward path at Site 2, both wandering methods give unacceptably low estimates. The reason for all these poor estimates is the great variation in plant density from one end to the other. The limit of 25 plants for the F and B paths implies that only a small proportion of the overall site is being sampled. Because the wandering-forward procedure collects data faster, it samples a smaller proportion than the wandering-quarters procedure and hence is liable to give less representative estimates. When the methods are not limited to 25 plants their estimates are reasonably accurate.

The results underline the need to sample *throughout* the region of interest.

Table 4.3 Estimates of abundance obtained using the wandering methods for the four mangrove sites illustrated in Figure 4.6. Point estimates within 20% of the true value are shown in bold. F and B refer to alternative 25-tree routes through the sites. The T routes traverse the sites from one end to the other.

	Site 1			Site 2			Site 3	Site 4
	F	B	T	F	B	T		
Wandering quarters, $\hat\rho_{w1}$, $\theta=\pi/4$	**434**	**540**	**628**	**1128**	606	**1016**	73	178
Wandering forward, $\hat\rho_{w1}$, $\theta=\pi/2$	1486	341	**528**	**1021**	690	**823**	88	308
Wandering forward, $\hat\rho_{w2}$	1703	**391**	596	**1170**	791	**926**	104	353
True value	472	472	472	990	990	990	85	227
No of trees sampled by wandering q.	25	25	35	25	25	102	8	16
No of trees sampled by wandering f.	25	25	79	25	25	174	10	25

4.10   Handling mixtures of species

Suppose a region consists of two competing types of plant, with one being common and the other rare. The object of an investigation is to assess the abundance of the rare plant. Two possible distance-based strategies are as follows:

1. Ignore the common plants and work only with distances from sampling points to rare plants.
2. Work with all the plants, assessing both the overall abundance and the proportion belonging to the rare type.

A problem for both strategies is that, for most of the sampling points, there may be no rare plants visible, or the nearest rare plant may be infeasibly distant. Since the chance of locating a rare plant increases when more plants are examined, this suggests using a method that involves the examination of many plants. The best performing of the methods examined in this chapter was conveniently one that involved many plants: $\widehat{\rho}_{(6)a}$.

Example 4.4: Mangroves in Kenya (cont.)

Sites 1 and 2, taken together, formed a complete section of the forest; from low tide shore to terrestrial vegetation. Site 1 contained four mangrove species: 122 *Avicennia marina* (A), 7 *Bruguiera gymnorrhiza* (B), 324 *Ceriops tagal* (C), and 19 *Xylocarpus granatum* (X). Their locations, which are illustrated separately in Figure 4.18, show considerable variations in intensity for each species.

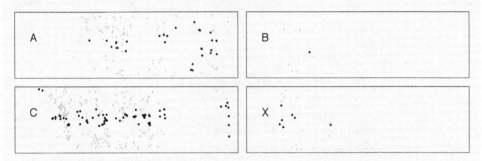

Figure 4.18 Site 1 contained four mangrove species (A, B, C, and X) with the locations indicated (the x-axis and the y-axis are on different scales). The larger symbols identify the locations of those mangroves selected when species is ignored in identifying nearest neighbours and $\widehat{\rho}_{(6)a}$ is used with a central row of sampling points.

Using $\widehat{\rho}_{(6)a}$ with 25 sampling points and the strategy of ignoring species when identifying the nearest neighbours, the data gatherer might use information from a total of 150 trees. In practice, however, it is likely that fewer trees will be identified, since some trees may be one of the six nearest neighbours for more than one sampling point. Using a central line of 25 sampling points in Site 1, it turns out that just 111 trees are identified. These trees are identified by using larger symbols in Figure 4.18.

Table 4.1 reported that, using $\widehat{\rho}_{(6)a}$, Site 1 was estimated as containing 461 trees. The upper part of Table 4.4 refers to this estimate. Since 29 of the 11 trees identified as part of the estimation process were of species A, the number estimated for the entire site is 461 * 29/111 = 120. The estimates for the common species are in excellent agreement with the true numbers. Those for the rarer species are uncertain, since small changes in the numbers identified will make a proportionally large change in the estimate. For example, if there had been two individuals of species B then the estimate would have doubled, whereas if the sampling had failed to include any individuals of that species then its presence would have passed unnoticed.

The second half of the table displays the results obtained if each species is separately analysed. The results are marginally better, but this is misleading. In a site with dimensions 16 m × 100 m, for some sample points, the nearest individuals of species B were more than 50 m distant and would not have been detected.

Table 4.4 The observed number, true number, and estimated number, using $\hat{\rho}_{(6)a}$, for the four mangrove species at Site 1. Estimates within 20% of the true value are shown in bold.

Species	A	B	C	X
Considering all species together				
Number used in estimation process	29	1	74	7
Estimated number based on proportion of overall estimate	**120**	4	**307**	29
Considering each species separately				
Number used in estimation process	65	7	93	16
Estimated number	**120**	10	**345**	17
Number actually present	122	7	324	19

4.11 Recommendations

The results reported in previous sections show that all the methods work well with random patterns, while most struggle when plant density is low, or when there are marked variations in plant density across the region of interest. The choice of method will depend upon what is practical in a particular context. A trial run may be helpful in making decisions about the method to be used. Some general observations follow.

- Sample points should be chosen so that the entire study area is covered, with roughly equal areas being 'allocated' to each point.
- If the results from individual points show marked variations in intensity across the study area, then a more accurate estimate is likely to be obtained by using the average of the separate estimates, rather than an estimate based on pooling information.
- Methods that involve measurements to relatively large numbers of neighbouring plants will usually give more accurate estimates than those based on single distances.
- Sheil's variant of the VAT procedure discussed in the next chapter is recommended if the methods in this chapter seem impractical.
- The best procedure appears to be to use $\hat{\rho}_{(k)a}$ with k as large as is practical.

5. Variable sized plots

5.1 Variable area transect (VAT)

Parker (1979) suggested a procedure that attempted to combine the speed of distance sampling with the accuracy of enumerating plants in a specified region. The procedure is illustrated in Figure 5.1.

The method uses n sampling points, with the observer walking in a specified direction from each point, until k plants have been located that are within a distance $w/2$ from the observer (see Figure 5.1). The estimator is

$$\hat{\rho}_{v(k)} = (kn - 1) \left/ \left(w \sum_{i=1}^{n} d_i \right) \right. , \qquad (5.1)$$

where d_i is the distance walked from the ith sampling point. For a random pattern the estimator has variance $\hat{\rho}_{v(k)}^2/(kn - 2)$. With k taken to be 3, Engeman et al. (1994) judged the VAT estimator to be one of the most practical. In a subsequent investigation, Engeman, Nielson and Sugihara (2005) concluded that a slightly larger value of k (say 5 to 7) was optimal. They also recommended that 'the transect be as wide as can be readily accommodated in a single pass'. In a follow-up investigation, comparing the VAT method with the most promising distance methods, White et al. (2008) concluded that 'the VAT method would seem the most straightforward to utilize in most field situations'. Dobrowski and Murphy (2006) suggest that the best choice for w is a value similar to the width of the objects being sampled.

However, when plant density varies noticeably across the study region, the approach of using the average of individual transect estimates, might be preferable. Denoting the n separate transect density estimates by $t_1, \ldots, t_n$, the estimate from sampling point i is given by

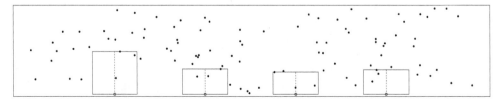

Figure 5.1 The variable area transect procedure: from each sampling point (marked with an o) the observer walks in a specified direction until k (here $k = 3$) plants have been observed within a transect of width w. The distance walked is recorded.

$$t_i = \frac{k-1}{wd_i}. \tag{5.2}$$

and the overall estimate is

$$\widehat{\rho}_{v(k)a} = \frac{1}{n}\sum_{i=1}^{n} t_i. \tag{5.3}$$

The standard deviation of the mean of the n separate estimates (which measures a combination of sample uncertainty and the variation in plant density) is

$$s = \sqrt{\frac{1}{n(n-1)}\left\{\sum_{i=1}^{n} t_i^2 - \frac{1}{n}\left(\sum_{i=1}^{n} t_i\right)^2\right\}}. \tag{5.4}$$

An approximate 95% confidence interval for the overall plant density is

$$(\widehat{\rho}_{v(k)a} - 2s, \ \widehat{\rho}_{v(k)a} + 2s).$$

Sheil et al. (2003), noted that 'in low density [plant] cover, [in order to obtain the required k plants] the sample may ultimately extend far from its origin, [and] cross vegetation and site types'. To avoid this, they defined three transect types with different formulae for t_i depending on the numbers of individuals present.

If no individuals have been observed by a minimum distance, $d_{\min}$, then the observer stops walking. This is a type 1 transect. If individuals are present, but scarce, then the observer walks the maximum distance, $d_{\max}$, and records k_{obs}, the number of individuals observed. This is a type 2 transect. If individuals are plentiful, then the distance walked to the kth plant is recorded. This is a type 3 transect.

The density estimates for the three transect types are given as the last column in Table 5.1. The overall density estimate, $\widehat{\rho}_{v(k)s}$, is the average of the separate transect estimates. If every transect were of type 3 then $\widehat{\rho}_{v(k)s}$ would equal $\widehat{\rho}_{v(k)a}$. As the required value of k is increased, so there will be fewer transects of type 3 and more of type 2. Thus, although increasing k reduces the variance for type 3 transects, the reduction of the number of this type of transect will increase the overall variability. A preliminary trial would be required to determine a suitable value for k.

Sheil's procedure is illustrated in Figure 5.2 for a case where $k=5$. Four empty transects are curtailed at $d_{\min}$. Four transects with $0 < j < 5$ are of length $d_{\max}$. The remaining two transects have lengths defined by the location of the fifth plant. In this case, with 51 plants in the study region, the separate transect estimates are 0, 0, 66.7, 66.7, 100, 0, 33.3, 263.2,

Table 5.1 The rules suggested by Sheil et al. (2003) for use with the VAT procedure when estimating plant density in heterogeneous regions. The minimum and maximum search distances from a sampling point are $d_{\min}$ and $d_{\max}$, respectively.

Type	No. of plants in $d_{\min}$	No. of plants in $d_{\max}$	Distance, d, to kth plant	Value for t_i
1	0			0
2	> 0	$k_{\text{obs}}(< k)$	$> d_{\max}$	$k_{\text{obs}}/wd_{\max}$
3	> 0	At least k	$< d_{\max}$	$(k-1)/wd$

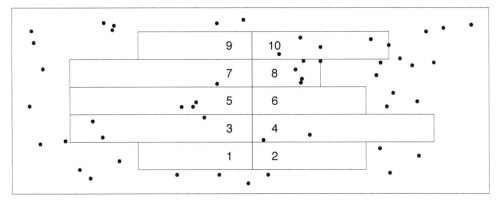

Figure 5.2 A compact example of Sheil's procedure. The empty transects 1, 2, 6, and 9 are truncated at d_{min}. Transects 8 and 10 are truncated at the fifth plant encountered. Transects 3, 4, 5, and 7, each containing from 1 to 4 plants, have length d_{max}.

0, and 178.6. These values give $\widehat{\rho}_{v(5)s} = 70.8$ and $s = 28.1$ so that the approximate 95% confidence interval is (15, 127). This is too wide to be useful and indicates the need for further sampling.

Figure 5.3 compares $\widehat{\rho}_{v(k)}$, $\widehat{\rho}_{v(k)a}$, and $\widehat{\rho}_{v(k)s}$ for the cases $k = 3$ and $k = 6$. The results are based on the usual 25 sampling points, with $w = 0.05$, $d_{min} = 0.15$, and $d_{max} = 0.2$. As expected, the results using $k = 6$ are less variable than those with $k = 3$. In clustered patterns $\widehat{\rho}_{v(k)a}$ tends to overestimate. This is a consequence of edge effects (sample points for which k plants have not been found are ignored using this method). Sheil's procedure appears to work well with every pattern type.

Figure 5.4 emphasizes the problems that may occur when searching for rare plants. For $\widehat{\rho}_{v(k)}$ and $\widehat{\rho}_{v(k)a}$ only a few sample points (in the relatively plant-rich regions) are providing usable information and therefore considerably over estimate. Although Sheil's method may give an inaccurate estimate, on average it does very well. The simplicity of the method suggests using more sampling points to reduce variability. Of course when plants are this scarce a complete census would be preferable.

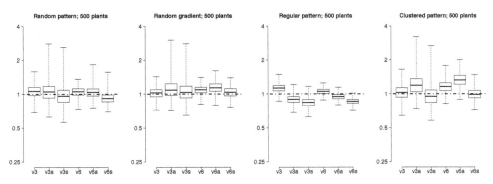

Figure 5.3 Box-whisker plots investigating the accuracy of the variable area transect methods when plants are abundant.

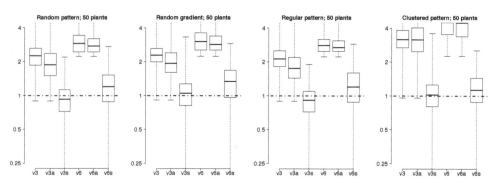

Figure 5.4 Box-whisker plots investigating the accuracy of the variable area transect methods when plants are scarce.

ⓘ Advice on data collection

For an Indonesian forest, Sheil (2002) suggested using transects of 10 m width, with $k = 5$, $d_{min} = 15$ m, and $d_{max} = 20$ m.

Example 5.1: Mangroves in Kenya (cont.)

Using the same sampling points as in Chapter 4, and noting that each sampling point must be visited, d_{max} was chosen to be the distance between successive sampling points (4 m for Sites 1, 3, and 4, but 3 m for Site 2). For each site, d_{min} was chosen to be $d_{max}/2$.

The results are summarized in Table 5.2. None of the estimators provide acceptable results for every site. With the given values of d_{min} and d_{max}, there is little difference between $\hat{\rho}_{v(k)a}$ and $\hat{\rho}_{v(k)s}$. The worst performance is that of $\hat{\rho}_{v(3)}$ for Site 1. None of the estimators achieve the accuracy found using the $\hat{\rho}_{(k)a}$ series given by Equation (4.6).

Table 5.2 The estimates obtained for the four mangrove sites using various VAT estimators with $k = 3$ or $k = 6$. Estimates within 20% of the true value are shown in bold.

Site	True no.	$\hat{\rho}_{v(3)}$	$\hat{\rho}_{v(3)a}$	$\hat{\rho}_{v(3)s}$	$\hat{\rho}_{v(6)}$	$\hat{\rho}_{v(6)a}$	$\hat{\rho}_{v(6)s}$
1	472	239	371	373	279	**394**	**401**
2	990	648	**1144**	**1145**	608	1191	1193
3	85	**88**	64	65	110	50	54
4	227	**196**	**200**	**201**	**218**	**198**	**200**

5.2   3P sampling

The phrase '3P sampling' is a shorthand for 'sampling with Probability Proportional to Prediction', a method of estimation introduced to the statistics community by Horvitz and Thompson (1952), and to forestry by Grosenbaugh (1964).

From a forest containing N trees, an initial sample of n trees is selected, with the aim of estimating some characteristic of interest. Most commonly this is V, the total volume of timber present. Let x denote some other characteristic that is highly correlated with the characteristic of interest. For volume, examples might be height (measured or estimated), basal area, or a visual estimate of the tree volume. Let x_{max} and x_{min} be the (possibly hypothetical) maximum and minimum values anticipated for this characteristic.

In a process sometimes referred to as *double sampling*, random numbers are used to select a subsample of the initial sample. Let r be a random number and let x_i be the value of the characteristic for the ith individual. The selection rule is that this individual is included in the second sample if

$$r < (x_i - x_{min})/(x_{max} - x_{min}).$$

This procedure implies that individuals with large x-values are more likely to be selected than those with small x-values. Since the next step is to accurately measure the characteristic of interest (in this case, volume), the implication is that the most important trees are given most attention.

Suppose that m of the initial sample of n trees are selected and denote the volume of the ith tree by v_i. The estimate of the total volume, V, is given by

$$\widehat{V} = \left(\sum_{i=1}^{N} x_i \right) \times \left(\sum_{i=1}^{m} \frac{v_i}{x_i - x_{min}} \right) \bigg/ \left(\sum_{i=1}^{m} \frac{x_i}{x_i - x_{min}} \right).$$

Usually the minimum value for x will be zero, in which case the expression simplifies to

$$\widehat{V} = \frac{1}{m} \left(\sum_{i=1}^{N} x_i \right) \times \left(\sum_{i=1}^{m} \frac{v_i}{x_i} \right). \tag{5.5}$$

Although primarily used to estimate volume of wood, 3P sampling has a far wider application. A useful discussion is provided by West (2011).

5.3   Bitterlich sampling

Most species of tree are characterized by having a trunk that can be regarded as a reasonably straight line that is either approximately vertical (if still growing), or approximately horizontal (if felled). Furthermore, trees tend to be in proportion: wider trees are taller and have greater volume. The methods in this section, and Section 5.4, exploit these characteristics.

The method introduced by Bitterlich (1948) is for use with standing trees. At each sampling point, the number of trees that have a trunk that subtends an angle greater than some pre-specified angle, θ, is recorded. Special instruments called *relascopes* are available to assist with the angle judgement. The method goes under several alternative names, including the *angle-count method*, and *point sampling*. The procedure may be regarded as a special case of 3P sampling (West, 2011).

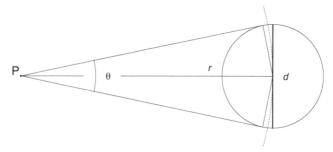

Figure 5.5 A tree of diameter *d* at a distance *r* from a sampling point subtending an angle *θ* at the sampling point P.

Suppose that r is the maximum distance at which a tree with basal diameter[1] d subtends an angle of θ or more (see Figure 5.5). To be included in the sample the tree must therefore be located somewhere in a 'catchment' area of πr^2 surrounding the sampling point. Since a tree of basal diameter d has a basal area of $\pi d^2/4$, the ratio of basal area to catchment area, β, is given by $d^2/4r^2$.

Now consider another tree, of diameter kd. This tree will subtend an angle θ or more, up to a distance kr from the sampling point (see Figure 5.6). Since this tree, which has basal area $\pi k^2 d^2/4$, is located somewhere in a catchment area $\pi k^2 r^2$, the ratio of basal area to catchment area is again $\beta = d^2/4r^2$.

For fixed θ, the quantity β, known to foresters as the *basal area factor* or *BAF*, will be the same for *all* distances. Suppose that, in a region of area A there are n sampling points, with t_i trees of a particular species subtending an angle of at least θ at sampling point i. Using only the information from sampling point i, the total basal area of the trees of that species in the region will be estimated by $A\beta t_i$. Since each sampling point gives an equally valid estimate, the pooled estimate of the total basal area is given by

$$B = \frac{A\beta}{n} \sum_{i=1}^{n} t_i = A\beta \bar{t}, \tag{5.6}$$

where $\bar{t}$ is the average number of trees counted at a sampling point. A confidence interval could be based on the variability of the individual estimates. An intriguing feature is that B is calculated without any actual measurements of individual trees.

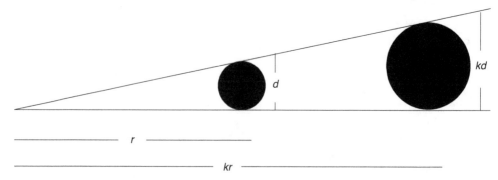

Figure 5.6 A tree of diameter *d* at a distance *r* from a sampling point subtends the same angle as a tree of diameter *kd* at a distance *kr* from the point.

ⓘ *Advice on data collection*

One question concerns the choice of value for θ. Too small a value would result in large numbers of eligible trees, some of which might be obscured by nearer vegetation. A recommended choice (Marshall, Iles and Bell, 2004) is a value for θ that results in about 4 to 10 trees being selected from each sampling point.

With a fixed-plot scheme, the catchment region leading to inclusion of a tree is the same for every tree, but with Bitterlich sampling the catchment area for a large tree is greater than that for a small tree (see Figure 5.7). For this reason the method is also called *variable-plot sampling* or sampling with *probability proportional to size*. Given that foresters refer to the transition between sampling points as a 'cruise', yet another description is *plotless cruising*.

Although an estimate of the total basal area can be obtained without making measurements on individual trees, this is not the case for volume, which requires measurement of basal area and height, together with the use of species-specific equations or look-up tables. Denoting the basal area of the jth tree at the ith sampling point by b_{ij}, with v_{ij} being the corresponding volume, the total volume of timber, V, is estimated by

$$\widehat{V} = \frac{1}{n} \sum_{i=1}^{n} \left(A\beta \sum_{j=1}^{t_i} \frac{v_{ij}}{b_{ij}} \right). \tag{5.7}$$

This is the average over the n sampling points of their separate estimates of total volume, with the information from each tree being inversely weighted according to the catchment area of the tree. In the forestry literature the ratio of volume to basal area is referred to as *VBAR*.

The estimated total basal area varies considerably from one sample point to another (reflecting, in part, the tendency for trees to appear in clusters). By contrast *VBAR* does

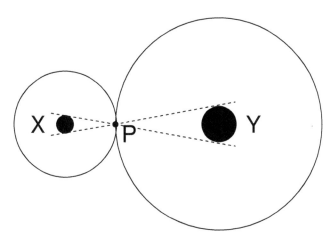

Figure 5.7 For tree X to be sampled, the sampling point must lie within the left-hand circle shown. Because tree Y is larger, the circle that surrounds it is larger. Thus Y has a greater probability of being selected by a randomly positioned sampling point.

not vary greatly from one tree to another. This suggests that, rather than using a relatively small number of sampling points, with every tree being individually measured, it will be more efficient to use a larger number of sampling points while only measuring a subsample of the trees encountered (an application of the 3P procedure of Section 5.2).

Using L as a subscript for the large sample and S as a subscript for the subsample, an estimate of volume, V_S, can be calculated using Equation (5.7) applied to the measurements of the trees in the subsample. Using Equation (5.6), two estimates of the basal area can be calculated: one from all the data, B_L, and one from the subset alone, B_S. To take account of the possibility that the subsample was slightly unrepresentative, an adjusted estimate of volume, $\widehat{V}_{adj}$ is then given by

$$\widehat{V}_{adj} = \frac{B_L}{B_S}\widehat{V}_S. \tag{5.8}$$

In order to select the subsample, Marshall, Iles and Bell (2004) suggested that, at each sampling point, in addition to selecting trees that subtended the angle θ, a second selection should be made using the larger angle Θ. This procedure is referred to as 'big BAF'. They suggested that the size of Θ should be chosen so that about 10 to 15 trees would be selected across the entire set of n_L sampling points. A detailed account that explores the interrelation between variability, cost and precision is provided by Yang et al. (2017).

5.4 Perpendicular distance sampling (PDS)

Williams and Gove (2003) adapted the ideas underlying Bitterlich sampling in order to estimate the total volume of 'coarse woody debris' (CWD) in an area of interest. CWD principally refers to fallen tree trunks, or logs of wood, with diameters greater than some minimum quantity (typically 7.5 cm).

When using PDS two measurements are required. One is the perpendicular distance, d, of the log from the sampling point (nearby logs are ignored if no perpendicular distance exists: see Figure 5.8). When estimating volume, the second measurement is s, the cross-sectional area of the log at the point where the perpendicular meets the log. The log is selected if d is less than cs, for some specified value of c (which has dimensions of length^{-1}). In effect, each log is surrounded by its own catchment region which has an area proportional to the log's volume. A log is selected only if the sampling point lies in its catchment. Figure 5.9 shows a typical catchment.

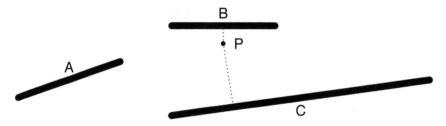

Figure 5.8 Three logs lie near P, a sampling point. Log A is not selected as no perpendicular distance exists. Whether log B and C are selected will depend on their cross-sectional areas at the feet of the respective perpendiculars.

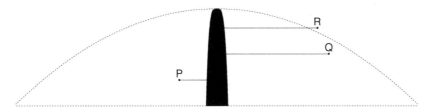

Figure 5.9 The shaded region represents a tapering log. The log will be selected only if the sampling point lies within the dotted catchment region. In this case the log would be selected by sampling points at P and Q, but not by a sampling point at R.

The size of the catchment area is directly proportional to the volume of the log. To see that this is so, suppose that log j is divided into N thin sections, each of thickness h/N. Suppose that the cross-sectional area of the ith section is s_i. The volume of the log, v_j is therefore given by

$$v_j = \sum_{i=1}^{N} s_i \frac{h}{N}.$$

Now consider the catchment area for this log. Since the cross-sectional area of the ith section is s_i, the width of the corresponding catchment section is cs_i on either side of this point of the log. The total catchment area, a_j is therefore given by

$$a_j = 2c \sum_{i=1}^{N} s_i \frac{h}{N},$$

which is directly proportional to the volume of the log since

$$a_j = 2cv_j.$$

The probability that, in a study region of size A, a randomly placed sampling point lies in the catchment for log j is $p_j = a_j/A$. Suppose that, for sampling point k, a total of m logs are selected. An unbiased estimate of the total volume is

$$\widehat{V_k} = \sum_{j=1}^{m} \frac{v_j}{p_j} = A \sum_{j=1}^{m} \frac{v_j}{a_j} = A \sum_{j=1}^{m} \frac{v_j}{2cv_j} = \frac{mA}{2c}. \tag{5.9}$$

With n sampling points the pooled estimate is given by

$$\widehat{V} = \frac{1}{n} \sum_{k=1}^{n} V_k. \tag{5.10}$$

A confidence interval for the overall volume would be based on the variability of the n separate estimates.

Williams, Ducey, and Gove (2005) adapted the original procedure to measure surface area (which is relevant since it may provide shelter or habitat for other organisms). The procedure is identical to that for volume, except that the critical measurement is the circumference of the log, rather than its cross-sectional area. From a sampling point, a log is counted if its perpendicular distance is less than some predetermined multiple, C, of the circumference. Note that, unlike c, C is a dimensionless quantity. For a sampling point that counts M logs, the total surface area is then, by analogy with Equation (5.9), estimated as $MA/2C$. An overall estimate of surface area is provided by the average of the n sample estimates, with its precision determined by the variance of the separate estimates.

ⓘ Advice on data collection

Using PDS there is no need to measure the diameters of every piece of CWD. Only borderline cases will need precise measurement. The values of c and C should be chosen so that between 4 and 10 logs are selected at most sampling points.

Gove et al. (2013) provide a comprehensive comparison of PDS methods (including the distance-limited version introduced in the next section).

5.4.1   Distance-limited PDS

One potential problem with the PDS procedure is that a relevant log might be a considerable distance from the sampling point. The answer suggested by Ducey et al. (2013) is to impose an upper limit, d_{max}, on that distance, as illustrated in Figure 5.10.

Define s_{max} by $s_{max} = d_{max}/c$, so that s_{max} is the cross-sectional area corresponding to d_{max}. The log is selected if either of two situations occur:

Case 1: $s \le s_{max}$ and $d \le cs$.
Case 2: $s > s_{max}$ and $d \le d_{max}$.

Suppose that at sampling point k, there are m_1 case 1 selections, and that the case 2 selections have cross-sectional areas $s_1, s_2, ..., s_{m2}$. For that sampling point the estimate of volume (Ducey et al., 2013) is

$$\widehat{V_k} = \frac{m_1 A}{2c} + \frac{A}{2d_{max}} \sum_{i=1}^{m_2} s_i, \tag{5.11}$$

with the overall estimate of volume being the average of the n separate estimates.

If s refers to circumference rather than area, and c is replaced by the dimensionless C, then the equivalent formula provides a distance-limited estimate of total surface area.

An alternative approach to the measurement of CWD was discussed earlier in Section 2.4.4. A comprehensive review of alternative methods is provided by Russell et al. (2015).

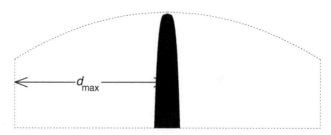

Figure 5.10 As previously, the shaded region represents a tapering log. With distance-limited PDS, the log will be selected only if the sampling point lies within the truncated dotted catchment region.

Part III

Mobile individuals

With moving objects there are two principal reasons why the number counted is likely to be less than the number present. One reason is simply that a creature, that might have been counted, is elsewhere in its territory at the time of counting. The other reason is simply that the creature is not seen! Birds are particularly problematical: there is the bird that flies behind the observer; the bird that dives under water just before the observer looks in that direction; the bird that vanishes into the foliage before it can be identified. Animals also cause difficulties: a small animal may be concealed behind a large bush, or behind the other creatures between it and the observer. Creatures under water also present obvious difficulties.

To estimate the number of moving targets it is therefore necessary to make assumptions. This requires mathematical models. Models will include unknowns that must be estimated. The estimation process is rarely simple, so the methods in this Part make relatively heavy use of computers and the examples include computer code.

6. Quadrats, transects, points, and lines – revisited

6.1 Box quadrats

The quadrats discussed at length in Chapter 2, were essentially two-dimensional, and not useful for a mobile individual that can 'escape' from the quadrat before being counted. For small ground-living creatures, one solution is to add vertical sides to the quadrat to form a 'box' that will trap the individuals of interest. The resulting data would be analysed in the same way as for ordinary quadrat data.

6.2 Strip transects

A strip transect is simply a narrow rectangular quadrat. Typically, the observer travels along the centre of the strip, recording all individuals as they are encountered. If it is believed that every individual is counted, then the methods discussed in Chapter 2 are appropriate. However, in most cases, the probability of detecting an individual reduces with increasing distance between the individual and the observer, so that the number recorded is less than the number present.

If, for each detected individual, the approximate distance from the observer to that individual is recorded, then the methods of Chapter 8 are appropriate.

If the distance is not recorded, but it can be assumed that the proportion of individuals that elude detection is the same from year to year (or site to site), then the numbers provide an index of relative abundance. This is the basis of the many examples of *citizen science*, where untrained observers record what they see on particular days or particular walks. Their aggregated findings can provide reliable year-to-year trends for the species concerned.

6.2.1 Bats

In the UK, the National Bat Monitoring programme was introduced in 1996. Volunteer walkers use ultrasonic detectors to detect bats by their calls. Depending on the target species, the transects are either triangular within a pre-specified randomly chosen 1 km square, or follow the path of a waterway. The transect counts could be used to provide an index of change.[1]

For North America, Britzke and Herzog (2009) suggested using *mobile (driving) transects* where the observer detects bats using a slow-moving car with an ultrasonic detector mounted on the roof. Their suggestions were incorporated into the protocol proposed by Loeb et al. (2015), which is applied to selected 10 km × 10 km quadrats across the continent. Key points are as follows:

> ### ❶ *Advice on data collection*
>
> - Drive slowly and steadily at 20 mph with headlights and hazard lights on.
> - Avoid main roads, gated roads, rough gravel roads, or roads through dense forest.
> - To avoid double counting, choose reasonably straight roads.
> - Using a 10 km × 10 km square grid for mapping, the route should remain within a single square.
> - Two surveys should be conducted within a fortnight during the bats' maternity season.
> - Surveys should begin 45 minutes after sunset on clear dry nights.

6.2.2   Marine fish

A trawl is in effect a line transect. Methods based on changes in the amount of fish caught are considered later in Section 6.7.2. Less invasive approaches to the measurement of fish abundance are introduced in Section 6.4.3.

6.2.3   Reef fish

Strip transects, in which a diver records what is seen, are the most common method used for sampling on coral reefs (Caldwell et al., 2016). For an *underwater visual census* (UVC), divers should follow a well-defined procedure, such as those described by Halford and Thompson (1994) and Labrosse, Kulbicki, and Ferraris (2002).

> ### ❶ *Advice on data collection*
>
> - Divers should work in pairs.
> - One end of a (typically) 50 m tape is fixed to the sea bed.
> - Before starting to record, allow fish time to acclimatize to the presence of the divers.
> - Proceed steadily along the tape, unwinding the tape and working on one short section (say 3 m) at a time.
>
> Halford and Thompson (1994) recommend counting the fastest moving species first; Labrosse, Kulbicki, and Ferraris (2002) advise counting the most abundant species first.

Remote underwater video (RUV) techniques avoid the disturbance caused by divers, and effectively give extra time for the observer. A comprehensive review of the methods used, showing the increased sophistication of the equipment available, is given by Mallet and Pelletier (2014).

Well-camouflaged species may be massively under-reported by an UVC. For example, a study by Willis (2001), showed that just 26 spectacled triplefin (*Ruanoho whero*) were observed at a location where 292 were present. Zarco-Perello and Enríquez (2019) showed that the use of RUV could more than double the number of species reported using UVC.

An alternative approach to underwater monitoring has been suggested by Widmer et al. (2019). Their scheme uses a series of small waterproof cameras, fixed at intervals in a line along the edge of a transect. All the cameras are similarly oriented and simultaneously take a sequence of pictures. so that the entire transect is being constantly monitored. The authors term the procedure *point-combination transect sampling*.

6.3 Using frequency to estimate abundance

In Chapter 2, it was shown that frequency (a record of either 0 for absent, or 1 for present) might be used to estimate abundance. Royle and Nichols (2003) suggested that a similar approach could be used with bird records, if, at each sampling point, a series of counts were made during a period in which the population size could be assumed to be constant. Each point count should have the same duration and refer to the same size region.

The underlying idea is simple. At a site where a species is abundant, individuals will be observed on most visits; where it is rare, it will be rarely observed. Thus variations in the probability of detection will reflect variations in abundance.

The model proposed by Royle and Nichols (2003) (hereafter the RN *model*) has two components: one concerns the number of birds present at a site, and the other concerns the number observed at a site. For example, if birds occur at random across a region, then the number present around sampling point i, N_i, will be an observation from a Poisson distribution (see Section 1.4.2). The parameter of the distribution, λ_i, will reflect the density of the species in the neighbourhood of the site. If all the sites are similar then $\lambda_i = \lambda$, for all i. If there are d distinct groups of sites, then there would be d different λ-values.

Suppose that, for each of the N_i individuals present at site i, the probability that it is detected on occasion j is p_{ij} independent of the detection of other individuals. The species will not be recorded at site i only if all N_i individuals remain undetected. The probability that the species is observed on visit j is therefore P_{ij} given by

$$P_{ij} = 1 - (1 - p_{ij})^{N_i}. \tag{6.1}$$

The value of the parameter λ_i may depend on the characteristics of the sampling site. The value of p_{ij} is likely to depend upon factors such as the weather, the time of day, the experience of the observer, and so forth. Models of this type have also been referred to as *abundance-induced heterogeneity (AIH) models*. The estimate of the overall abundance may be obtained by scaling up from the area of the sampling site to the larger area of the region being sampled.

Fiske and Chandler (2011) presented the versatile R package *unmarked* which can be used to fit many abundance models including the RN model described above. To be sure of obtaining valid parameter estimates, the estimation procedure works with $\log(\lambda)$ and $\log(p_{ij}/(1 - p_{ij}))$, so that some manipulation is required to make sense of the results obtained.

Example 6.1: Wood Thrushes in New Hampshire

Royle and Nichols (2003) examined the fit of the RN model to several sets of data including a set of observations of Wood Thrush (*Hylocichla mustelina*) at 50 sites in New Hampshire. A total of 11 visits were made to each site during a period of 30 days in 1991 as part of the North American Breeding Bird Survey. The data for the Wood Thrushes are provided as part of the *unmarked* package.

Table 6.1 (a) Summary of the numbers of occasions (out of 11) on which Wood Thrushes were observed at 50 sites. (b) Summary of the numbers of sites (out of 50) on which Wood Thrushes were observed on the 11 visits.

(a) Number of occasions	0	1	2	3	4	5	6	7	8	9	10	11
Number of sites		5	8	7	11	2	1	1	4	4	1	3	3

(b) Visit	1	2	3	4	5	6	7	8	9	10	11
Number of sites	12	16	11	17	21	21	23	21	22	22	20

Table 6.1 summarizes the survey results. Part (a) shows that there were three sites where Wood Thrushes were always detected and five where it was never noted. Part (b) suggests that the numbers increased in the later part of the observation period.

Using *unmarked* the analysis begins with the rather unlikely model in which the probability of observing a Wood Thrush is the same at each site (constant λ), and the same at each sampling period (constant p).

```
library(unmarked)            # Calls the appropriate R library

#woodthrush.bin is the 50  11 matrix of 0s and 1s
# that correspond to 'not detected' and 'detected'

woodthrushUMF <- unmarkedFrameOccu(woodthrush.bin)
occuRN(~ 1 ~ 1, woodthrushUMF)
```

This gives the following output extract:

```
Abundance:
  Estimate    SE    z  P(>|z|)
     0.792 0.158 5.03    5e-07

Detection:
  Estimate    SE    z  P(>|z|)
    -1.21 0.17 -7.14 9.41e-13

AIC: 633.9534
```

The estimate of λ is exp(0.792) = 2.21 and the estimate of p is

$$\exp(-1.21)/(1 + \exp(-1.21)) = 0.23.$$

This is the start of the analysis, not the end! Table 6.1 (b) showed that the numbers of sites where wood thrushes were detected was much higher during the final seven visits than during the first four. This might reflect a change in behaviour of the birds brought on by increased visibility as a result of the need to feed young. The analysis could proceed as follows:

```
earlylate<-factor(c(1,1,1,1,2,2,2,2,2,2,2))
occuRN(~ earlylate ~1, woodthrushUMF)
```

The resulting output shows an unchanged estimate of abundance since the higher frequencies in the later visits are supposed to be a result of easier detection rather than more birds. The output concerning the probability of detection now has two entries:

```
Detection:
               Estimate    SE     z  P(>|z|)
(Intercept)    -1.63 0.205 -7.98 1.52e-15
earlylate2      0.65 0.181  3.58 3.39e-04

AIC: 622.4348
```

The first entry, (intercept), refers to the first value of the vector earlylate and gives an estimate of p as $\exp(-1.63)/(1 + \exp(-1.263)) = 0.16$. The second entry refers to the difference between the two periods. Since $-1.63 + 0.65 = -0.98$, the probability of detection in the second period is estimated to have increased to $\exp(-0.98)/(1 + \exp(-0.98)) = 0.27$. The lower AIC value for this model suggests that this model should be preferred to its predecessor.

In reality, the 50 sampling sites are unlikely to be equally suitable for Wood Thrushes. If there is information about the sites, then it is possible to model the effect of these background factors. To illustrate the approach, suppose that sites 1 to 12 are near water, and that sites 11, 12, and 41 to 50 have many trees. Then the analysis would continue:

```
water<-factor(c(rep(1, times=12),rep(2,times=38)))
trees<-factor(c(rep(1, times=10),2,2,rep(1, times=28),
rep(2,times=10)))
occuRN(~ earlylate ~ water+trees, woodthrushUMF)
occuRN(~ earlylate ~ water, woodthrushUMF)
```

The vector water assigns level 1 to sites near water. The trees vector assigns level 1 to sites without many trees and level 2 to sites with trees. The model is a great improvement, since the AIC value is much reduced:

```
Abundance:
               Estimate    SE     z  P(>|z|)
(Intercept)     2.349 0.263  8.93 4.13e-19
water2         -1.391 0.205 -6.79 1.15e-11
trees2         -0.924 0.308 -3.00 2.70e-03
```

```
Detection:
            Estimate    SE     z  P(>|z|)
(Intercept)   -2.153 0.271 -7.94 1.97e-15
earlylate2     0.607 0.178  3.41 6.55e-04

AIC: 581.5511
```

This model considers four types of site, each with its own estimate of abundance. For sites without trees, but near water: $\exp(2.349) = 10.5$. For sites without trees and not near water: $\exp(2.349 - 1.391) = 2.6$. For sites with trees and near water: $\exp(2.349 - 0.924) = 4.2$. Finally, for sites with trees but not near water: $\exp(2.349 - 1.391 - 0.924) = 1.0$.

Both tail probabilities (the entries in the $P(>|z|)$ column are very small suggesting that both background factors are relevant. The much smaller tail probability associated with proximity to water would indicate that this was the dominant feature.[2]

Since each point count refers to an area of $\pi/16$ square miles, multiplying the λ-values by $16/\pi$ would give the estimated density per square mile.

6.4 Point counts (point transects)

Point-count sampling generally requires one or more observers to go to a specified point, and report what is there. It is not expected that a point count will detect all individuals present, but its magnitude will depend upon the conditions under which it is made. For example, a novice observer, with poor sight, counting birds in eclipse plumage on a misty morning, is likely to count fewer birds than a sharp-sighted experienced observer counting the same birds in their breeding plumage on a sunny day.

Also, if twice as many individuals of species A are detected, compared to the number for species B, then this does not imply that species A is twice as abundant as species B. That may be true, but it may also be true that species A is more easily detected.

However, a comparison of the counts for a particular species made on different years can provide a valid *index* of change in population size, provided that the counts are made under comparable conditions. Comparability is also likely if the counts have been aggregated across many sampling points, as is the case for national surveys.

6.4.1 Birds

In order for the information from different observers at different locations to be comparable, there needs to be a clear set of instructions that are followed by all. Ralph, Droege, and Sauer (1995) (henceforth RDS) set out a list of 28 recommendations that covered all aspects of the counting process. However, when Matsuoka et al. (2014) (henceforth M) studied 20 years of bird count data from Canada and Alaska, they found that less than 3% of the point counts, had exactly followed the recommendations concerning the length of the count, and the count radius. They endorsed the earlier recommendations, which are given below, together with some elaborations suggested by M.

ⓘ *Advice on data collection*

- **Count radius:** Infinite. RDS specified that the numbers observed within 50 m, and the numbers recorded at greater distances, should both be recorded. M suggested that the ideal would be exact distances (so that the methods of Chapter 8 could be employed). Failing that, they suggest using at least four distance ranges, with 50 m being one of the end points.
- **Count period:** RDS suggesting counting for either 5 or 10 minutes. Numbers should be recorded separately for the periods 0–3 minutes, and 3–5 minutes (and 5–10 minutes, if the longer period is used). M suggested subdividing the 5–10-minute interval into two, with the division at 8 minutes. M suggested that consideration might be given to recording subsequent detections of a bird previously detected (so that the methods of Chapter 7 might be used).
- **Number of observers:** RDS had in mind a single observer. However, using two observers allows the methods of Section 6.5 to be used.
- **Number of visits:** RDS anticipated single visits per season, but multiple visits may give improved estimates. Thus, MacLeod et al. (2012), found that 'A single five-minute count had c. 60% chance of detecting bellbirds [*Anthornis melanura*] at a location where they were present, while the cumulative detection probability increased to almost one after five repeat counts per survey.'

Note that one visit is likely to be sufficient if distance methods (Chapter 8) are used.

The recommendations by RDS were put together so as to allow some degree of comparability across different studies, but, as M noted, there are considerable variations. Thus the North American Breeding Bird Survey (BBS), which has been running since 1966, uses 3-minute counts. The New Zealand forest counts in the 1980s were for 5 minutes (Hartley, 2012). Gibbons and Gregory (2006) recommend normally using 5 or 10 minutes, with longer counts in forests.

However, the use of an extended time interval may lead to overestimation, since some individuals may be double-counted and others may be included that were simply passing through the region. For this reason, Smith et al. (1998) recommended a maximum duration of 10 minutes.

The arrival of the observer at the counting point may scare individuals away, so it is advisable to avoid under-estimation by waiting for a fixed time before commencing the count. Gibbons and Gregory (2006) suggest one minute for birds.

For the BBS, each observer is assigned a 24.5 mile route, with sampling points at either end, and at half-mile intervals between. There are more than 4000 such routes, with start points and travel directions chosen at random, with the aim of providing an even covering across the entirety of North America. In most cases the routes follow minor roads, so that the sampler will not be disturbed by traffic noise. Each route is sampled once a year at the height of the breeding season. The guidelines given above are not precisely followed, since an upper distance bound of 400 m is used. Here are some more general BBS instructions:

> **ℹ** *Advice on data collection*
> - Avoid sampling on a day with poor visibility, rain, or strong winds.
> - Count during the early morning.
> - A full 360° scan should be used.
> - Keep track of the locations of individuals, noting if several are simultaneously present.
> - It may help to record positions on paper, but avoid looking down at your notes for extended periods.

Buckland, Marsden, and Green (2008) warned that

> In many bird monitoring surveys, no attempt is made to estimate bird densities or abundance. Instead, counts of one form or another are made, and these are assumed to correlate with bird density. Unless complete counts on sample plots are feasible, this approach can easily lead to false conclusions, because detectability of birds varies by species, habitat, observer and many other factors. Trends in time of counts often reflect trends in detectability, rather than trends in abundance. Conclusions are further compromised when surveys are conducted at unrepresentative sites.

Because of these problems, they advise using distance methods (see Chapter 8). Their criticism is supported by the results of the seven-year study reported by Norvell, Howe, and Parrish (2003).

6.4.2   Butterflies

The North American Butterfly Association has co-ordinated point counts since 1993. It requires teams of volunteer observers to count all butterflies seen within a single day, within circles of radius 7.5 miles around pre-specified points. Comparisons of the results across years, are used to monitor changes in butterfly populations, and to study the effects of weather and habitat change.

6.4.3   Remote underwater surveying

According to the manual produced by Johannesson and Mitson (1983), the first scientific record of the use of an acoustic method for detecting fish was a Japanese publication in 1929. A very readable account of the subsequent development of acoustics for the assessment of the abundance and ecology of fish, and other marine life, is provided by Fernandes et al. (2002). The quantity measured is amount of fish, rather than number of fish.

Underwater photography can provide more detailed information. In an overview of the use of videos of fish coming to bait, Cappo, Harvey, and Shortis (2006) claim that 'The use of remote, baited "video fishing" techniques offer standardized, non-extractive, methodologies for estimating relative abundance of a range of marine vertebrates and invertebrates, with the option of very precise and accurate length and biomass estimates when stereo-camera pairs are used.' Harvey et al. (2007) assessed the number of fish present by determining the value of *maxN*, defined as the maximum number of fish seen at any one time. Despite its name, maxN provides a lower bound on the number of fish present. Denney et al. (2017) suggest that improved accuracy is gained by using a rotating camera. Marini et al. (2018) report promising results using genetic programming to automate the counting of fish in video footage. They find that the most important requirement is to avoid bio-fouling of the camera.

6.4.4   Camera traps

A camera trap is a remotely activated camera, equipped with an appropriate sensor, that collects a series of images of creatures that pass. Burton et al. (2015) provide a comprehensive review of previous studies that used camera traps. Niedballa et al. (2016) introduced the R package *camtrapR* which provides management of camera trap data, and an interface to relevant R packages.

Camera trap data are often summarized using a *relative abundance index* (RAI), which may be calculated either as the proportion of photographs containing the species of interest, the average number of individuals per photograph, or the number of distinct animals (if the creatures in question are distinctively marked).

A camera situated by a known trail will be more successful than one in a thicket, while a camera with high resolution and wide range will be more successful than a poor quality camera with a narrow range of vision. Therefore, comparisons of two sets of RAI values will only be reasonably reliable, if the equipment and sampling procedures are comparable. However, for a permanent array of cameras, variations in the numbers of individuals observed from year to year, should provide information concerning changes in abundance.

6.4.4.1   *Trapping region*

The region effectively scanned by a camera depends upon its distance above ground and the size of the targeted species. However, the regions scanned by identical cameras set at identical heights may still differ. This is because there may be variations in the inclination of the local terrain relative to the camera, and also in the amount of cover in the vicinity of the camera.

Rowcliffe et al. (2008) proposed a *random encounter model* where the creatures of interest were supposedly distributed at random, and moving in random directions. They showed that the estimated density, λ, was then given by

$$\lambda = \frac{\alpha\pi}{vr(2+\theta)}, \tag{6.2}$$

where α is the number of pictures taken by a camera in unit time, v is the speed of the target, and r and θ define the camera detection zone. In practice, individuals are likely to be moving along well-worn paths, but the model is reasonable providing that the cameras have not been positioned to specifically cover these paths.

If individuals are moving in groups, then the total number present would be estimated as

$$\lambda \bar{g} A, \tag{6.3}$$

where $\bar{g}$ is the average group size and A is the area of the target region.

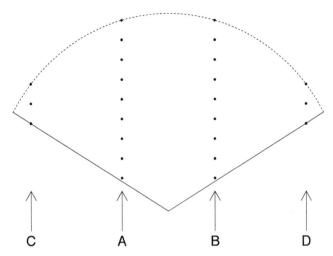

C   A   B   D

Figure 6.1 Detection of a passing creature depends on its speed, direction and proximity to the camera. It also depends upon the frame speed of the camera. The diagram indicates the paths of four individuals. Those at locations A and B are more likely to be detected than those on the parallel paths at C and D. Similar results occur whatever the paths.

Figure 6.1 shows that a target is more likely to be detected if it passes near the centre of the field of view because, in effect, the camera has a longer time in which to detect the individual's presence.

The following example suggests that the effective field of view of a camera may be best determined after a study of the results.

Example 6.2: Bawean warty pigs

Rademaker et al. (2016) conducted a study of the warty pig *Sus blouchi* on Bawean island, Indonesia. Their data, which are available at the public repository http://data.4tu.nl, provide information concerning the positions of pigs when first sighted.

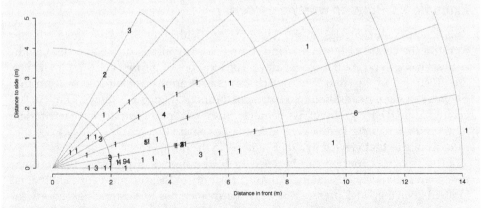

Figure 6.2 The locations of groups of pigs relative to the camera when first detected. The results for all cameras are superimposed. Numbers are group sizes.

Figure 6.2 superimposes the results from 34 cameras, with sightings at angles up to 60° and distances up to 14 m. The figure suggests that the region covered by the cameras may not be a simple sector: at 60° the most distant pig recorded was 5 m away; at 25° the most distant pig recorded was about 9 m away; the three most distant records refer to pigs noted at angles of 10° or less. A reasonable approximation for the catchment region for these pigs might be a rectangle of dimensions 10 m × 15 m, with an area of 150 m².[3] Rowcliffe et al. (2011) examined camera trap records of agoutis (*Dasyprocta punctata*) and also found 'a relative lack of records with both large radius and large angle'.

6.4.4.2   Using the RN model

Section 6.3 introduced the RN model of Royle and Nichols (2003). The model assumes that the individuals of interest are randomly distributed across the region of interest with a density of λ per unit area. At each sampling point, the model requires there to be a sequence of independent samples. The time period over which sampling occurs must be sufficiently short that it can be assumed that the local population is unchanged. Crucially, the model also assumes that no individual is observed at more than one sampling point.

In the present application each 'sampling point' is a camera. What is not so straightforward is the definition of a sample. This must be a section of time during which the camera either records (1), or fails to record (0), an individual of interest. Too small a time interval will give too few 1s. Too long an interval will give too few 0s. The problem is equivalent to that faced when choosing the quadrat size to determine plant frequency (Section 2.5). There is also the question of the size of the gap between successive records. If there is a zero gap, then double counting must be avoided for individuals that remain in view across the time boundary between intervals. If there is a non-zero gap, then there is the danger that valuable data will have been ignored.

The example that follows demonstrates the overestimates that may arise if several cameras monitor the same territory.

Example 6.3: Bawean warty pigs (cont.)

This example concentrates on the data obtained within the 32.6 km² Blok Gunung forest in the centre of Bawean island. Cameras were installed five at a time, with each set being in position for seven days. The locations of the first eight deployments are illustrated in Figure 6.3, with each camera set approximately 300 m from the previous camera. In addition to the camera locations, Figure 6.3 shows 1 km circles centred around each group of five cameras. Since wild pigs typically have territories of around 4 km, a pig family viewed from one camera in a group is likely to be the same family as that viewed by other cameras in the group.

The *unmarked* routine can be used to fit the model in precisely the fashion described previously in Section 6.3. If the proximity of the cameras to one another is ignored, then the 40 × 7 presence/absence matrix includes 19 separate sightings with an average group size of 1.7. The model output included:

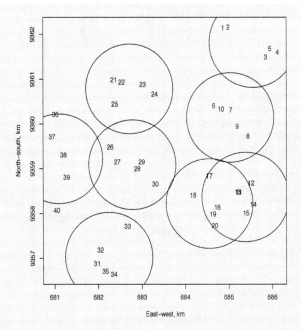

Figure 6.3 The locations of the first eight deployments of five cameras. The circles have radius 1 km and are centred on the midpoints of each group of five cameras.

```
Abundance:
Estimate   SE    z P(>|z|)
  0.466 1.04 0.45   0.653
```

Assuming that the (very imprecise) estimate for $\log(\lambda)$ relates to a region of 150 m^2, this corresponds to the infeasible estimate of around 11,000 pigs per km^2!

A more reasonable approach is to regard each set of five cameras as a single 'observer', with each observer covering about 3 km^2.[4] The basic data is now an 8 × 7 array, with a total of 15 sightings. The resulting output is

```
Abundance:
Estimate   SE    z P(>|z|)
  2.55 0.896 2.85 0.00435
```

The estimate of abundance is now about 4 pigs per km^2, with an approximate 95% confidence interval from 2 to 75 km^2. Given the overlap between deployments 3 and 4, the estimate is still likely to be an overestimate. Using Equation (6.2), Rademaker et al. (2016) gave an estimate of around 6 per km^2.

6.5 Double-observer sampling

There are at least three variants of what is recorded when there are two observers at a sampling site.

6.5.1 Dependent double-observer sampling (DDS)

Suppose that it is known that, for every individual present, the probability of being observed is p. Then, if n individuals were observed, the estimated number present would be n/p. Nichols et al. (2000) presented a method adapted from one proposed by Cook and Jacobson (1979) for estimating bias in aerial surveys. The method, which uses two observers, results in two counts being recorded for each sampling point. The first count is the number of individuals observed by the 'primary' recorder; the second count is the number of individuals observed by the 'secondary' recorder that were not observed by the primary recorder. During a sequence of point counts, the two observers should alternate the primary and secondary roles.

ⓘ *Advice on data collection*

In the context of bird populations, Forcey and Anderson (2002) made a number of recommendations that included

- 'Unlimited radius counts produce the greatest number of detections and are useful for recording rare species',
- 'Longer counts [10 minutes] reduce observer bias because investigators have more time to record individual birds and ensures data are accurate'

Let p_i ($i = 1, 2$) be the detection probability for observer i. The probability that observer j sees an individual not seen by observer i is therefore $(1 - p_i)p_j$. Let n_{ij} be the number of individuals (totalled over all sampling sites) counted by observer j when observer i is the primary observer. Thus, for a site with M individuals, on average

$$n_{11} = Mp_1 \quad \text{and} \quad n_{12} = M(1 - p_1)p_2. \tag{6.4}$$

Thus, on average, $n_{12}/n_{11} = (1 - p_1)p_2/p_1$. Similarly $n_{21}/n_{22} = (1 - p_2)p_1/p_2$. Solving these simultaneous equations gives estimates of the detection probabilities for the two observers as

$$\widehat{p_1} = \frac{n_{11}n_{22} - n_{12}n_{21}}{n_{22}n_{10}} \quad \text{and} \quad \widehat{p_2} = \frac{n_{11}n_{22} - n_{12}n_{21}}{n_{11}n_{20}}, \tag{6.5}$$

where $n_{i0} = n_{i1} + n_{i2}$ (for $i = 1, 2$). The probability that an individual is observed by at least one observer is $p = 1 - (1 - p_1)(1 - p_2)$. Using Equations (6.5) this is estimated by

$$\widehat{p} = 1 - \frac{n_{12}n_{21}}{n_{11}n_{22}}. \tag{6.6}$$

The total number of individuals seen across all the sampling points is $n_{00} = n_{10} + n_{20}$, so that the total number of individuals present is given by

$$\widehat{N} = n_{00}/\widehat{p}.$$

Formulae for the variances of the estimated probabilities were given by Cook and Jacobson (1979). The estimated variance for $\widehat{p}_1$ is given by

$$\widehat{\text{Var}(\widehat{p}_1)} = \frac{\widehat{p}_1(1-\widehat{p})(n_{00}-\widehat{p}n_{10})}{\widehat{p}_2(1-\widehat{p}_2)n_{10}n_{20}}, \tag{6.7}$$

with a corresponding result for $\widehat{\text{Var}(\widehat{p}_2)}$. Also

$$\widehat{\text{Var}(\widehat{p})} = (1-\widehat{p})^2\widehat{p}\left\{\frac{1}{\widehat{p}_1 n_{10}} + \frac{1}{\widehat{p}_2 n_{20}} + \frac{1}{\widehat{p}_2(1-\widehat{p}_1)n_{10}} + \frac{1}{\widehat{p}_1(1-\widehat{p}_2)n_{20}}\right\}, \tag{6.8}$$

Nichols et al. (2000) showed that

$$\widehat{\text{Var}(\widehat{N})} = \frac{n_{00}^2\widehat{\text{Var}(\widehat{p})}}{\widehat{p}^4} + \frac{n_{00}(1-\widehat{p})}{\widehat{p}^2} \tag{6.9}$$

The programme *DOBSERV* (available with an R version) implements their approach.

The methods in this section are most effective when there are many sampling points, as in the North American Breeding Bird Survey. However, they have been used in other situations as the following example illustrates.

Example 6.4: Alpine marmots

Corlatti et al. (2016) compared four methods for estimating the population size of Alpine marmot (*Marmota marmota*) in a study area of the Stelvio National Park in Italy. To test double-observer sampling they used four sites in the summer of 2016 obtaining the results (primary, secondary): (8, 0), (2, 0), (2, 1), (1, 1). The two observers alternated roles, so that $n_{11} = 10$, $n_{22} = 3$, $n_{12} = 1$, and $n_{21} = 1$. Thus $n_{00} = 15$, $p_1 = 29/33 = 0.88$, $p_2 = 29/40 = 0.725$, and $p = 1 - 1/30$ giving the estimated number of marmots as $15 \times 30/29 \approx 15.51 \approx 16$. This is very low, since capture-recapture methods (discussed in Chapter 7) showed that at least 54 marmots were present in the study area.[5]

6.5.1.1   Modelling site-to-site variability

Royle (2004) assumed that the numbers of individuals present at equal-sized sampling sites varied according to an underlying Poisson distribution (Section 1.4.2) with parameter λ. The R package *unmarked* of Fiske and Chandler (2011) can be used to fit the model and assess the relevance of background information concerning the sites (for example, in the case of birds, the value of λ might vary according to the amount of foliage present). Comparing the AIC values (Section 1.9) of alternative models provides information concerning the relevance of the background variables included in the analysis.

Example 6.5: Alpine marmots (cont.)

Using *unmarked* the following commands produce the results for the model that ignores the differences between observers (denoted here as a and b):

```
library(unmarked);
Obs<-as.data.frame(matrix(c('a','b','a','b'),4,1))
mdata<-matrix(c(8,2,2,1,0,0,1,1),4,2)
mframe<-unmarkedFrameMPois(mdata,Obs,type='removal')
model1<-multinomPois(~1~1,mframe);model1
```

An extract from the resulting output is:

```
Abundance:
Estimate    SE    z  P(>|z|)
     1.35 0.261 5.16 2.47e-07

AIC: 30.10935
```

The model estimates the value of $\ln(\lambda)$ as 1.345704 which is reported in the extract as 1.35 and corresponds to an estimate for λ as 3.84. This is the estimated number of individuals per sample site and corresponds to an overall estimate of 15.36 for the sampled region. The next step is to take account of possible differences between the observers:

```
model2<-multinomPois(~1~Obs,mframe);model2
```

This gives the following output extract:

```
Abundance:
              Estimate    SE     z  P(>|z|)
(Intercept)     1.73 0.304  5.69 1.26e-08
Obsb           -1.01 0.584 -1.73 8.32e-02

AIC: 28.71239
```

Taking account of the observers results in an improved model (a lower AIC value). The average number of marmots per site observed when observer a is the primary observer is exp(1.73) = 5.63, and that when observer b is the primary observer is exp(1.73 – 1.01) = 2.05, once again giving an overall total of 15.36.[6]

The difference between the results of the observers is significant (the programme gives a low tail probability: p = 0.0083) though this may reflect differences in the sampled sections, rather than differences between the observers. Since each individual had only two opportunities to be primary observer, it would be unwise to draw any firm conclusion.

With just four sampled sites there is considerable uncertainty in any parameter estimate. For example, the command

```
exp(confint(model1,type='state'))
```

gives the 95% confidence interval for the mean number of marmots per sampling point:

```
             0.025     0.975
lambda(Int) 2.303711 6.403775
```

Summing over four sampling points, would give a 95% confidence interval for the total in the region sampled, as between 9.2 and 25.6. Note that the lower value is less than the number observed (15).

6.5.2 Independent double-observer sampling (IDS)

In this case, at each sampling site, the two observers independently record where and when individuals are observed. At the end of each sampling session, the observers compare notes, so that three counts are produced as summarized in Table 6.2.

Table 6.2 Notation for counts and the corresponding expected numbers when a population of size N is sampled by observers a and b. The probabilities of the two observers noticing an individual present at the sampling site are p_a and p_b.

	Number observed	Expected number observed
Observed by a only	n_a	$Np_a(1 - p_b)$
Observed by b only	n_b	$Np_b(1 - p_a)$
Observed by both a and b	n_{ab}	Np_ap_b

Fletcher and Hutto (2006) recommend IDS for 'river bird surveys, particularly for small-bodied species that have low detection probabilities'. However, Forcey et al. (2006) preferred DDS to IDS in the context of general bird surveys. For the detection of bat species, Duchamp et al. (2006) found that using two non-overlapping remote bat locators could provide a significant improvement over the use of a single locator.

Equating the observed counts (totalled over all the sampling sites) with the corresponding expected numbers gives an estimate of p_a as $n_{ab}/(n_{ab} + n_b)$, with a corresponding result for p_b. The estimated number of individuals present is given by

$$\widehat{N} = \frac{(n_{ab} + n_a)(n_{ab} + n_b)}{n_{ab}}. \tag{6.10}$$

The R package *unmarked* of Fiske and Chandler (2011) can again be used to estimate N together with a confidence interval for the population size.

Example 6.6: Alpine marmots (cont.)

Table 6.3 Numbers of marmots observed by two observers, a and b, during the summer of 2015 in four sectors (A, B, C, D) in Italy's Stelvio National Park.

Sampling point	A	B	C	D	Total
Observed by a only	2	1	0	1	$n_a = 4$
Observed by b only	0	2	1	4	$n_b = 7$
Observed by both a and b	5	0	3	2	$n_{ab} = 10$

Table 6.3 shows the results obtained viewing marmots using the IDS approach. The estimated number using Equation (6.10) is

$$\widehat{N} = 14 \times 17/10 = 23.8.$$

Using *unmarked* the relevant R commands are:

```
mdata2<-matrix(c(2,1,0,1,0,2,1,4,5,0,3,2),4,3)
mframe2<-unmarkedFrameMPois(mdata2, type='double')
exp(confint(multinomPois(~1~1,mframe2),type='state'))
```

The output (which refers to the mean of the underlying Poisson distribution of sampling point counts) is:

```
              0.025     0.975
lambda(Int) 3.805484 9.479751
```

Summing over the four sampling points, this corresponds to a 95% confidence interval for the total number of marmots present as (15.2, 37.9). Note that, as for the DDS example, the lower bound is less than the number of marmots actually observed (21).

6.5.3   Unreconciled double-observer sampling (UDS)

A logistical difficulty with both DDS and IDS is that the two observers talk to one another! This takes time, can distract from the process of observation, and can lead to bias if one observer influences the other. Riddle, Pollock, and Simons (2010) suggested using the much simpler double-observer alternative in which the two observers make independent counts. Their comparison of UDS with IDS suggested that IDS gave more precise estimates but was also more prone to bias.

Let the number of individuals present at sampling point i be N_i. Using the model of Royle (2004), N_i is assumed to be an observation from a Poisson distribution with parameter λ. The value of λ might vary from one type of site to another (e.g. heavily wooded as opposed to open grassland), but is assumed to be constant for all sampling points of the same type. Each observer is presumed to have a constant probability of observing each individual at a site (though this might vary from one type of site to another).

Example 6.7: Alpine marmots (cont.)

From Table 6.2 it is apparent that observer a observed 7, 1, 3, and 3 marmots from the four sampling points, while the counts for observer b were 5, 2, 4, and 6. Combining the maximum counts from each sampling point shows that there were at least 7 + 2 + 4 + 6 = 19 marmots present. Using the *unmarked* package once again, the commands

```
library(unmarked);
mdata2a<-matrix(c(7,1,3,3,5,2,4,6),4,2)
model3<-pcount(~1~1,mframe2a,K=50);
exp(coef(model3)[1])
exp(confint(model3,type='state'))
```

lead to the output

```
  lam(Int)
 7.500442
            0.025    0.975
 lam(Int) 2.1913 25.67272
```

In this case the overall estimated number of marmots is higher than previously: 4 × 7.5 = 30. However, it is still a considerable under-estimate, since Corlatti et al. (2017) report that there were at least 54 marmots present. The 95% confidence interval of (8.8, 102.7) is very wide and indicates the need for a great deal of further sampling.

6.6 Double sampling

Each of the DDS, IDS and UDS methods uses distributional assumptions to estimate detection probabilities to arrive at estimates of abundance. Bart and Earnst (2002) proposed a more direct approach. They suggested using a large number (M) of sampling points, with some quick sampling procedure being used at each. At a small number (r) of these sampling points (randomly chosen) a thorough investigation should be carried out to discover the true number of individuals present. If the number reported at these sites was n_r, while the true number was found to be N_r, then the probability of detection is estimated as $\hat{p} = n_r/N_r$. If the total number observed over all M sampling points was n_M, then the true number would be estimated as $n_M/\hat{p}$. Bart et al. (2004) observe that an advantage of double sampling is that it produces unbiased estimates.

6.7 Removal sampling

In removal sampling the individuals observed in each sample are excluded from consideration by subsequent samples. One method is to physically remove them from the population. A less radical option is to mark them in some fashion, so that, in subsequent samples, it is known that they have already been observed. The counts of interest are the numbers of new individuals caught in each successive sample.

6.7.1 The regression approach

Suppose that the target population (of unknown size, N) is sampled using k trapping periods. The critical assumptions are that every individual has the same probability, p, of being trapped, that p is the same for every trapping period, and that the population is closed (so that all the individuals are present throughout the k trapping periods and no new individuals arrive.) Suppose that n_i individuals are first observed during the ith trapping period, and let the number previously observed be x_i, so that

$$x_i = n_1 + n_2 + \cdots n_{i-1} \qquad (i = 2, 3, \ldots, k), \text{ with } x_1 = 0. \qquad (6.11)$$

With these assumptions Leslie and Davis (1939) observed that e_i, the expected number of new individuals trapped in the ith period is given by

$$e_i = Np - px_i. \qquad (6.12)$$

They suggested estimating p and Np by using the observed counts $\{n_i\}$ as the values for $\{e_i\}$ and applying a standard linear regression routine to the pairs of (x_i, n_i) values. A potential problem with this approach is that the estimated value for N can be less than the total number actually trapped.

If there are just two sampling periods, then the estimates $\widehat{N}$ given by and $\widehat{p}$ are

$$\widehat{N} = \frac{n_1^2}{n_1 - n_2} \qquad \text{and} \qquad \widehat{p} = \frac{n_1 - n_2}{n_1}. \qquad (6.13)$$

These estimators only exist if $n_2 < n_1$.

Example 6.8: Okaloosa darters in Florida

Dorazio, Jelks, and Jordan (2005) provide data concerning the Okaloosa darter (*Etheostoma okaloosae*), an endangered species of freshwater fish. Distinct subsections of stream crossings at Eglin Air Force Base in Florida were examined in the same way on each occasion, with all individuals observed being temporarily removed. Data for one crossing are given in Table 6.4, together with the corresponding fitted numbers based on Equation (6.12).

Table 6.4 The numbers of Okaloosa darters removed in one subsection of a stream crossing in Florida, together with the numbers previously removed.

Period	1	2	3	4	5
Numbers of darters removed, n_i	24	11	9	8	1
Numbers previously removed, x_i	0	24	35	44	52

Figure 6.4 plots the (x_i, n_i) pairs together with the fitted regression line. The estimated value for λ is –0.4034 and the estimate of the intercept, λN is 23.1061. Thus the estimated value for N is 23.1061/0.4034 = 57.3.

If only the data from the first two trapping periods are used then, using Equation (6.13), the estimated number of darters is $24^2/(24 - 11) = 44.3$, which is fewer than the number eventually trapped.

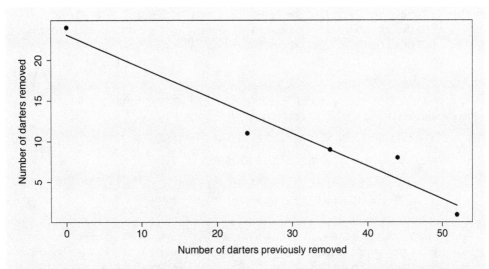

Figure 6.4 The numbers of darters caught plotted against the number previously caught. The fitted regression line is also shown.

6.7.2 Catch per unit effort (CPUE)

In the previous section, it was assumed that the process of removal is the same on each occasion. Suppose instead that the amount of effort put into the removal of individuals varies across the removals. Let f_i be the effort used (measured in some appropriate fashion) for the ith removal, and let the probability of an individual being trapped be directly proportional to the trapping effort:

$$p_i = \lambda f_i. \tag{6.14}$$

Following the previous argument, e_i, the expected number of new individuals captured is given by

$$e_i = N\lambda f_i - \lambda f_i x_i. \tag{6.15}$$

Now define c_i, the *catch per unit effort (CPUE)* by

$$c_i = \frac{e_i}{f_i}, \tag{6.16}$$

Dividing through Equation (6.15) by f_i, gives the linear regression relation

$$c_i = N\lambda - \lambda x_i, \tag{6.17}$$

from which estimates of λ and N are easily obtained.

CPUE is easily measured and so is often used as a measure of abundance when monitoring stocks of fish. Since the regression approach avoids the use of the binomial distribution, it is not necessary to know the actual numbers caught. Instead, x_i may be any suitable measure of the catch size. In the context of fisheries, this will usually be the total weight caught. A detailed discussion of the reliability of CPUE as an index of population size is provided by Maunder et al. (2006).

In effect, the assumption being made is that, at time i, $c_i = qN_i$, where N_i is the size of the available population, and the constant, q, may be referred to as the *catchability coefficient*. In this context, Harley, Myers, and Dunn (2001) examined data from trawl surveys, and fitted the more general model

$$c_i = qN_i^\beta. \tag{6.18}$$

They found that for many species $\beta < 1$, indicating that the size of the population would be overestimated if the simple linear relation were used.

Example 6.9: Prince Edward Island lobsters

DeLury (1947) provided data (an extract is given in Table 6.5) concerning the capture of lobsters off Prince Edward Island in 1944.

Table 6.5 Weights of lobsters caught in traps off Prince Edward Island during May and June 1944. Catch is total weight given in hundreds of pounds to the nearest hundred. Effort is the number of traps given in hundreds to the nearest hundred.

Catch	28	69	77	53	88	63	36	81	81	83	84	68	12
Effort	38	72	88	58	95	67	37	82	86	91	91	79	12
Catch	15	112	70	59	32	63	30	63	56	30	46	47	36
Effort	15	116	85	78	34	80	47	81	80	52	71	86	69

Figure 6.5 (a) plots the total weight of lobsters against the number of lobster pots. Figure 6.5 (b) plots CPUE against weight of lobsters previously caught. Since the CPUE value is approximately constant for the first 13 sampling points illustrated, these points provides no useful information concerning population size. The second group of 13 points do show the anticipated trend and fitting the linear relation of Equation (6.17) to these points gives $c = 1.66 - 0.00077x$. This line is illustrated in Figure 6.5 and implies that there were about 215,000 pounds of lobsters available at the beginning of the second half of the sampling period.

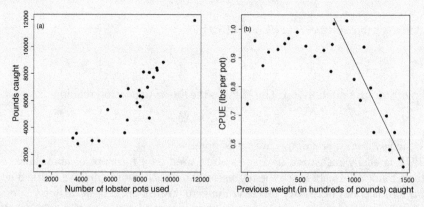

Figure 6.5 (a) Weight (in lbs) of lobsters caught plotted against the number of lobster pots used. (b) CPUE plotted against the weight of lobsters previously caught. The best fit line to the last 13 points is shown.

6.7.3 The likelihood approach

This method, introduced by Moran (1951), is sure to lead to an estimate of N that is at least as large as the number observed. The target population is sampled using k equal length trapping periods. The critical assumptions are that every individual has the same probability, p, of being trapped and that p is the same for every trapping period. Suppose that initially there are N individuals in the population, and that n_i individuals are removed from the population during the ith trapping period. With k trapping periods the log(likelihood) of this outcome is L given by

$$L = \log\{\mathrm{P}(N, p; n_1)\} + \log\{\mathrm{P}(N - x_2, p; n_2)\} + \cdots + \log\{\mathrm{P}(N - x_k, p; n_k)\}, \quad (6.19)$$

where $\mathrm{P}(N, p, r)$ is the binomial probability of obtaining r successes in N trials, with the probability of a success being p (see Equation (1.5)). In general a maximization algorithm (such as optim in R) is required to find the values $\widehat{N}$ and $\widehat{p}$ that maximize L and give the maximum likelihood estimates and a confidence region (see Section 1.6.4).

Example 6.10: Okaloosa darters in Florida (cont.)

Using the complete data, maximization of the likelihood given by Equation (6.19) gives estimates of $\widehat{N} = 56.2$ and $\widehat{p} = 0.42$, with fitted counts of 23.6, 13.7, 7.9, 4.6, 2.7. In this case the likelihood and regression methods give very similar results.

Figure 6.6 gives a contour plot of the likelihood with various confidence regions. Although the maximum likelihood estimate is that there are about 56 darters present in the region sampled, the 95% region suggests that the true number might be over 70.

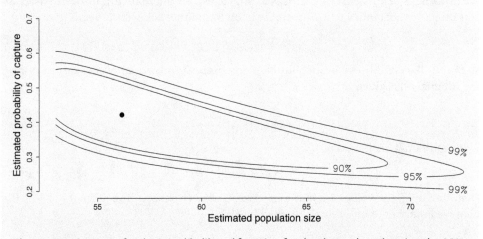

Figure 6.6 Contours for the joint likelihood function for the darter data showing the 90%, 95%, and 99% confidence bounds, together with the location of the maximum.

6.7.4 Unequal-length sampling intervals

Farnsworth et al. (2002) suggested a version of removal sampling that uses different-length sampling periods. Their motivation was that the point bird counts used for the North American Breeding Bird Survey typically result from 10-minute recording periods divided into successive intervals of lengths 3 minutes, 2 minutes, and 5 minutes, with the numbers first seen in each interval being separately recorded.

Suppose, more generally, there are k time intervals, with interval i having length t_i units. Let q be the probability of an individual not being observed during a unit of time (assumed to be the same for all individuals and all time). Thus q^t is the probability that an individual is not observed during an interval of length t while $(1 - q^t)$ is the probability that it is observed.

For an individual to be detected for the first time in the second interval, that individual must have been undetected in the first time interval, but detected in the second interval. Denoting the number of individuals present by N, and the probability of an individual being detected for the first time in interval i by p_i:

$$p_1 = 1 - q^{t_1}, \tag{6.20}$$

$$p_i = q^{T_i}(1 - q^{t_i}), \qquad (i = 2, 3, \dots, k), \tag{6.21}$$

where T_i is the total detection time prior to the ith detection period:

$$T_i = t_1 + \cdots + t_{i-1} \qquad (i = 2, 3, \dots, k). \tag{6.22}$$

The joint likelihood, L, is again given by Equation (6.19), but with these probabilities replacing the previously constant p. Although estimates of N and q can be obtained by maximizing L, Farnsworth et al. (2002) suggested an alternative approach based on conditional probabilities. The probability of an individual being observed is given by

$$P = 1 - q^{T_k + t_k} \tag{6.23}$$

so that c_i, the conditional probability that an observed individual was observed during the ith interval is given by

$$c_i = \frac{p_i}{P}. \tag{6.24}$$

The conditional likelihood is

$$C = c_1 \times c_2 \times \cdots \times c_k. \tag{6.25}$$

Denoting by $\hat{q}$, the value of q that maximizes C, the corresponding estimate of N is the observed number of individuals divided by $(1 - \hat{q})$.

Example 6.11: Ovenbirds in the Great Smoky Mountains

Farnsworth et al. (2002) give summarized results for many bird species. The counts given in Table 6.6 are the total counts of ovenbirds (*Seiurus aurocapilla*) made at 155 sampling points in the Great Smoky Mountains National Park on the border between North Carolina and Tennessee. Birds were only counted if they were within 50 m of the observer.

Table 6.6 The numbers (totalled over 155 sample points) of Ovenbirds observed (for the first time) in three successive recording intervals, together with the fitted values resulting from maximizing the conditional likelihood.

	First period (2 minutes)	Second period (3 minutes)	Final period (5 minutes)
Observed	141	29	39
Fitted (based on $q = 0.722$)	151	36	25

The following R code provides the estimate for q:

```
likelihood=function(q){ p1<-((1-q^3)/(1-q^10))^141
p2<-(q^3*(1-q^2)/(1-q^10))^29
p3<-(q^5*(1-q^5)/(1-q^10))^39
p1*p2*p3 }
optimize(likelihood,c(0,1),maximum=TRUE)
```

This code gives the estimated value of q as 0.722. Since there were a total of 209 ovenbirds observed, the estimated number present is $209/(1 - 0.722^{10}) = 217.3$. Thus almost all the birds present had been observed by the end of the 10-minute sampling periods. These counts are totals of observations at 155 sampling points, each monitoring an area of 2500π m², so the estimated density of ovenbirds is 1.79 per hectare.

6.7.5 Change-in-ratio method

This method relies on the population consisting of two types of individual whose relative frequencies are materially altered by the removal of individuals between two sampling periods. It was originally proposed by Kelker (1940) in the context of deer hunting.

At time i ($i = 1$ or 2), let N_i and n_i be the population and sample sizes. Let X_i be the numbers of x-type individuals in the populations at time i, with x_i being the number in the sample at that time. Suppose that, between times 1 and 2, n_r individuals are removed (or added), with x_r being the number of x-type individuals. Equating the proportions in the samples to the population proportions at those times gives two simultaneous equations with the resulting estimate of the initial population size being given by

$$\widehat{N_1} = \frac{x_2 n_r - x_r n_2}{x_2 n_1 - x_1 n_2} n_1. \tag{6.26}$$

Replacing the final n_1 by x_1 gives the estimated number of x-type individuals at the time of the first sample. An equivalent result expressed in terms of the proportions of x-individuals is

$$\widehat{N_1} = \frac{x_r - p_2 n_r}{p_1 - p_2}. \tag{6.27}$$

The method can only be expected to provide reliable results if the numbers observed are very large and the proportions observed differ markedly.

Example 6.12: Newfoundland snow crabs

Dawe, Hoenig, and Xu (1993) present information concerning snow crabs (*Chionoecetes opilio*) caught in St Mary's Bay Newfoundland in the fishing season between August and October 1991. Crabs were divided into classes according to their carapace width. Crabs were referred to as *x*-type if they had a carapace width of at least 95 mm (the minimum legal size). Crabs removed during the fishing season are therefore predominately *x*-type. Crabs were trapped both before and after the fishing season, using both small-mesh and large-mesh traps. Since the smaller crabs will pass through the large-mesh traps, the proportion of crabs that are *x*-type will be greater for these traps. The 1991 results are summarized in Table 6.7.

Table 6.7 The proportions of *x*-type crabs in samples before and after the fishing season, together with the proportion observed among those caught during the fishing season.

	Trap size	Number	Proportion *x*-type
Before the fishing season	Large-mesh		0.5425
	Small-mesh		0.3407
Fishing season		687,949*	0.9134
After the fishing season	Large-mesh		0.4893
	Small-mesh		0.2703

* The number of crabs caught during the season is estimated from the total catch weight.

Using Equation (6.27) with the large-mesh results gives

$$\widehat{N_1} = \frac{(0.9134 - 0.4893) \times 687949}{0.5425 - 0.4893} \approx 5.5 \text{ million.}$$

For comparison, the small-mesh results give

$$\widehat{N_1} = \frac{(0.9134 - 0.2703) \times 687949}{0.3407 - 0.2703} \approx 6.25 \text{ million.}$$

Although Dawe, Hoenig, and Xu (1993) present (rather complex) expressions for the variances of the estimates, in view of the indirect method of assessing the number caught, it would seem appropriate to state that, in the area being fished there were initially between 5 and 7 million snow crabs. It would appear that the population was decimated during the fishing season.

7. Capture-recapture methods

Capture-recapture methods are also called *mark-recapture methods*, or *capture-mark-recapture methods*.

A simple example is the following:

1. A sample of individuals is obtained from a population.
2. The individuals are marked in some way (possibly by the addition of identifying tags) and returned to the population.
3. A second sample is taken from the population.
4. The individuals in the second sample are examined to determine the proportion that are marked.
5. Taking this as an estimate of the proportion of marked individuals in the entire population, the population size can be estimated.

Early researchers using the method assumed that the population contained the same individuals throughout the sampling process with no immigration, emigration, births, or deaths. This assumption defines a *closed population*. When the samples are taken in quick succession, for example on successive days of a week, the assumption is a reasonable one. Later workers proposed models that included gains and losses of individuals. The population is then called an *open population*.

If the marked individuals are individually identifiable (for example via numbered tags) then the analysis can take into account their capture histories. Otherwise, the analysis simply relies on the numbers recaptured.

There is a huge literature on the application and analysis of capture-mark-recapture data. In this chapter, which can give no more than a brief introduction, the mathematical background behind the variety of distinct approaches will be largely avoided, with only the simpler formulae provided. In some situations there are no explicit formulae, with the computer being required to choose parameter values to maximize some complex expression. This is not a problem, since there is a wealth of freely available computer packages (a survey is provided by Bunge, 2013). Note, however, that, where the packages are based on different theoretical approaches, or make different assumptions or approximations, the output from one package may not match that from another. Unfortunately, there is no single approach that is optimal in all situations (Grimm, Gruber, and Henle, 2014).

An essential assumption for all methods is that there is not a hidden subpopulation of astute individuals that always avoid capture!

7.1 Capture-recapture models for a closed population

7.1.1 Two trapping occasions

The first use of capture-mark-recapture methods was made by Petersen (1896), who wished to estimate the size of a fish population. The method gained prominence, following its use by Lincoln (1930) to estimate the size of a bird population. Nowadays, the method that they used is variously described as the *Petersen mark-recapture estimator*, the *Lincoln index*, the *capture-recapture method*, or by some variant of these.

On the first occasion, from a population of unknown size, N, a total of n_1 individuals are captured, marked, and returned to the population. On the second occasion n_2 individuals are trapped and examined. Suppose that m_2 of these individuals are found to have been marked. Equating the proportion of marked individuals on the second occasion (m_2/n_2) to the proportion of marked individuals in the entire population (n_1/N) suggests that an estimate of the population size is given by:

$$\widehat{N_P} = \frac{n_2}{m_2} n_1. \qquad (7.1)$$

The method assumes that every individual in the population had the same probability of being trapped on each occasion, which implies that individuals trapped on the first occasion do not become more wary (or less wary) as a consequence of that experience. Chapman (1951) showed that $\widehat{N_P}$ is a biased estimator of N and suggested reducing the bias by using

$$\widehat{N_C} = \frac{(n_2 + 1)(n_1 + 1)}{m_2 + 1} - 1. \qquad (7.2)$$

Sadinle (2009) studied the accuracy of confidence intervals based on estimators similar to $\widehat{N_C}$. Using the notation of Table 7.1, Sadinle (2009) found that the most accurate confidence interval was given by

$$n_{11} + n_{12} + n_{21} - c + \widehat{n_{22}} \exp\left(\pm z \sqrt{\frac{1}{n_{11} + c} + \frac{1}{n_{12} + c} + \frac{1}{n_{21} + c} + \frac{1}{\widehat{n_{22}}}}\right), \qquad (7.3)$$

where

$$\widehat{n_{22}} = \frac{(n_{12} + c)(n_{21} + c)}{n_{11} + c}, \qquad (7.4)$$

$c = 0.5$, and, for an approximate 95% confidence interval, $z = 1.96$.

Table 7.1 Two-sample capture-recapture data presented as a 2×2 table. Entries in brackets are the unobserved numbers of interest.

		Sample 2		
		Captured	Not captured	Total
Sample 1	Captured	n_{11}	n_{12}	n_1
	Not captured	n_{21}	(n_{22})	
	Total	n_2		(N)

Example 7.1: New Zealand skinks

Wilson et al. (2017) provide a detailed analysis of capture-recapture data concerning the incidence of skinks (*Oligosoma*) at Macraes Flat, Otago. The data reported in this example refer to captures in 2009, within a region that was free from mammals, and protected by a mammal-resistant fence. There were 107 skinks captured on the first occasion, and 188 captured on the second occasion. Table 7.2 gives more detail.

Table 7.2 The numbers of skink captures and recaptures for two samples in a fenced plot in Otago, New Zealand.

		Sample 2		
		Captured	Not captured	Total
Sample 1	Captured	18	89	107
	Not captured	170		
	Total	188		?

The Chapman estimate is

$$\widehat{N_C} = \frac{(188+1) \times (107+1)}{18+1} - 1 \approx 1073.$$

Also $\widehat{n_{22}}$ is given by

$$\frac{170.5 \times 89.5}{18.5} = 824.85$$

so that an approximate 95% confidence interval is

$$276.5 + 824.85 \times \exp\left(\pm 2\sqrt{\frac{1}{18.5} + \frac{1}{89.5} + \frac{1}{170.5} + \frac{1}{824.85}}\right) \approx (758, 1689).$$

There were three different species of skink captured at Otago. The separate estimates for each species, the totals of those estimates, and the previous results are summarized in Table 7.3.

Table 7.3 The numbers of skink captures and recaptures for 3 species in a fenced plot in Otago, New Zealand, together with the population estimates and approximate 95% confidence intervals.

	Capture numbers			Population	
	n_1	n_2	m_2	$\widehat{N}_C$	95% c. i.
McCann's skink (*Oligosoma maccanni*)	69	114	11	670	(442, 1197)
Southern grass skink (*O. aff. polychroma* clade 5)	30	58	6	260	(161, 576)
Cryptic skink (*O. inconspicuum*)	8	16	1	76	(35, 509)
Sum of the above	107	188	18	1006	
All skinks, ignoring species	107	188	18	1073	(758, 1689)

Notice that the sum of the separate estimates is not equal to the overall estimate and that the confidence intervals are not centred on their respective $\widehat{N_C}$ values.

Small samples inevitably give rise to less precise results with wider confidence intervals. Narrower confidence intervals would be obtained by taking the overall confidence interval and reducing it in proportion for each species. In total there were

$$107 + 188 - 18 = 277$$

skinks observed. Of these,

$$8 + 16 - 1 = 23$$

were cryptic skinks. This suggests a population estimate of

$$1073 \times \frac{23}{277} \approx 89$$

cryptic skinks, with confidence limits of

$$758 \times \frac{23}{277} \approx 63 \qquad \text{and} \qquad 1689 \times \frac{23}{277} \approx 140.$$

The interval width is reduced from 474 to 77. An implicit assumption is that cryptic skinks are just as likely to be captured and recaptured as other skink species.

7.1.2 Alternative models with multiple trapping occasions

The Petersen model assumed that all the individuals in a population behave in the same fashion. A more plausible assumption would be that some individuals are more likely to be captured than others (the latter, perhaps, being older and wiser). This model, having heterogeneous capture probabilities, will be denoted by M_h.

Since it is also likely that the probability of recapture will vary as a consequence of changes in time of day, weather conditions, etc., a further model of interest is one that allows for the capture probabilities to vary from one occasion to another. This is model M_t. A model that allows for both heterogeneity and time dependence is denoted as M_{th}.

These model types all assume that the experience of being captured has no bearing on whether an individual is subsequently recaptured. A model that allows for the behaviour of an individual to be affected by a previous capture will be denoted by M_b. Since the individuals may also display varying capture probabilities, and since time effects may still play a part, there are several other model types to consider. These models will be denoted by M_{bh}, M_{tb} and M_{tbh}.

For completeness, there is also the somewhat unrealistic model M_0. This model excludes all these effects, by asserting that all individuals have the same probability of being captured, and that this is not affected by either their past capture history or the prevailing conditions for capture.

The connections between the eight model types are illustrated by the model tree shown in Figure 7.1. Each line connects a pair of models that differ by a single type of effect (time variation, behavioural response, or heterogeneity). Note that parallel lines indicate model pairs that differ in the same fashion, thus M_h and M_{th} differ by the addition of a time effect; this is equally true for M_{bh} and M_{tbh}.

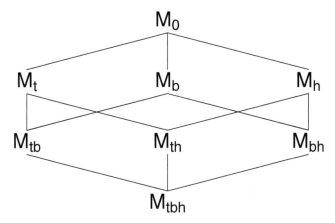

Figure 7.1 The tree of possible model types for capture-recapture data.

7.1.2.1 Presenting outcomes

Table 7.4 summarizes the skink results for the first three capture occasions. The data are presented as an incomplete three-way cross-classification. Because there is no obvious pattern to the frequencies there is no obvious value to replace the '?'.

An undesirable feature of Table 7.4 is that it is not easy to follow the fortunes of individuals. This would be still more of a problem if there were further capture occasions. A simpler procedure is to denote each possible set of outcomes by a sequence of 1s and 0s, where 1 signifies that the individual was captured and 0 signifies that it was not captured. Thus the sequence 01001 indicates that there were five trapping occasions, with the individual being captured on the second occasion and recaptured on the fifth occasion.

Table 7.4 Summary table of the results of the first three skink captures.

	Caught (3rd)		Not caught (3rd)	
	Caught (2nd)	Not (2nd)	Caught (2nd)	Not (2nd)
Caught (1st)	10	4	8	85
Not caught (1st)	24	82	146	?

Example 7.2: New Zealand skinks (cont.)

In 2009, the Otago skink capture-recapture experiment took place over 5 days. A total of 426 different skinks were caught in the fenced plot. Their capture histories (ignoring species) are summarized in Table 7.5. Of particular interest is the unknown number that were never caught (00000 in the table).

Table 7.5 The skink captures (ignoring species) on five successive days in a fenced plot in Otago, New Zealand.

Capture history	00000	00001	00010	00011	00100	00101	00110
Frequency	?	36	27	4	70	8	2
Capture history	00111	01000	01001	01010	01011	01100	01101
Frequency	2	121	13	11	1	6	4
Capture history	01110	01111	10000	10001	10010	10100	11000
Frequency	5	9	78	2	5	4	3
Capture history	11001	11010	11011	11100	11101	11110	11111
Frequency	1	3	1	1	3	2	4

7.1.3 *Modelling the capture probability

An extended discussion of the nature of the model types, together with an introduction to alternative statistical formulations for the models, is provided by Chao (2001). The sections that follow attempt to describe just a few of the associated ideas.

The previous section introduced the factors that might affect the probability of an individual being captured, but gave no details. In this section alternative model structures are introduced in the context of model M_{tbh}. A simple idea is that individual i has an intrinsic probability of being captured (h_i, say). At time j this is inflated or deflated by a factor t_j, with the same t factor applying to all individuals. In the same way, the probability of capture at time j, is inflated or deflated by a factor b_{ij} that is the same for all individuals that have the same capture history as individual i. Putting these factors together gives

$$P(\text{Individual } i \text{ captured at time } j) = h_i t_j b_{ij}. \tag{7.5}$$

Since parameter estimation routines are well established for linear models it is convenient to take (natural) logarithms to give:

$$\ln(P(\text{Individual } i \text{ captured at time } j)) = \ln(h_i) + \ln(t_j) + \ln(b_{ij}). \tag{7.6}$$

A problem with Equations (7.5) and (7.6) is that, when the estimated values of the unknown parameters are substituted back into the equations, the estimated value for P(Individual i captured at time j) might exceed 1. To avoid this problem (and because of other considerations affecting the estimation routines), rather than estimating P(Individual i captured at time j) it is convenient to work with the *logit*:

$$L_{ij} = \log\left(\frac{P(\text{Individual } i \text{ captured at time } j)}{P(\text{Individual } i \text{ not captured at time } j)}\right). \tag{7.7}$$

The resulting so-called *logistic model* may then be written as

$$L_{ij} = \alpha_i + \beta_j + \gamma_{ij}, \tag{7.8}$$

where α_i, β_j and γ_{ij} capture the individual, time, and historical effects. Exponentiating and rearranging this model gives

$$P(\text{Individual } i \text{ captured at time } j) = \frac{\exp(\alpha_i + \beta_j + \gamma_{ij})}{1 + \exp(\alpha_i + \beta_j + \gamma_{ij})}. \quad (7.9)$$

Logistic models were first applied to capture-recapture data by Pollock, Hines, and Nichols (1984). Their application in the present context is due to Huggins (1989), who observed that α_i might be a function of factors such as age and weight, while β_j might be a function of factors such as temperature and rainfall.

Except in the simplest cases, analysis of capture-recapture data requires the use of a computer programme. For closed populations, commonly used programmes are *CAPTURE* introduced by Otis et al. (1978), and *MARK* introduced by White and Burnham (1999). *MARK* is not limited to the analysis of closed populations, but can be used for any of the situations described in this chapter. There is a freely downloadable book (more than 800 pages of exemplary text together with another 300 pages of appendices) which describes every situation with far more detail than is given here.

7.1.4 Model M_0 (A single invariant capture probability)

This model supposes that the probability that an individual is captured, is a constant that is unchanging over time, is the same for every member of the population, and is unaffected by whether the individual has been previously captured. These are somewhat unlikely assumptions, but the model provides a useful yardstick for assessing the improvements obtained when one or more of these constraints is relaxed.

Darroch and Ratcliff (1980), assuming that capture histories were known, suggested using

$$\widehat{N_{DR}} = \frac{TF}{T - f_1}, \quad (7.10)$$

where T is the total number of captures, F is the number of *different* individuals captured, and f_1 is the number of individuals captured once only.

Example 7.3: New Zealand skinks (cont.)

Over the 5 capture occasions there were $T = 579$ skink captures. These involved $F = 426$ different skinks, with 332 skinks being caught on just one occasion. Thus $\widehat{N_{DR}} = 579 \times 426/(579 - 332) \approx 999$. This would imply that fewer than half the skinks present were caught during these five sampling occasions.

7.1.5 Model M_t (Time variation)

Like model M_0, this model assumes that, on each trapping occasion, every individual has the same probability of being captured. However, the model allows the common capture probability to vary from one time to another. This would be likely to be true if the climatic conditions varied from one capture time to another, or if there were unequal-length capture periods (as used by observers for the annual Breeding Bird Survey in North America). For this latter context see Alldredge et al. (2007).

Schnabel (1938) extended the Petersen mark-recapture estimator for use with t trapping occasions. With an extension to the previous notation, suppose that there are

m_i marked individuals among the n_i individuals trapped on the ith occasion, and that the total number of marked individuals at that time is M_i (so that, in particular, $M_2 = n_1$). Schnabel suggested using

$$\sum_{i=2}^{t} n_i M_i \Big/ \sum_{i=2}^{t} m_i \, ,$$

which is a weighted average of Petersen estimates and does not require the individual capture histories. To reduce the bias in small samples, Chapman (1954) suggested using the modification

$$\widehat{N_S} = \sum_{i=2}^{t} n_i M_i \Big/ \left(1 + \sum_{i=2}^{t} m_i\right). \tag{7.11}$$

An approximate 95% confidence interval based on an assumed normal distribution for $1/\widehat{N_S}$ is

$$\left(\frac{\widehat{N_S}}{1 + 2S\widehat{N_S}}, \qquad \frac{\widehat{N_S}}{1 - 2S\widehat{N_S}}\right), \tag{7.12}$$

where

$$S = \sqrt{\sum_{i=2}^{t} m_i \Big/ \sum_{i=2}^{t} n_i M_i} \, . \tag{7.13}$$

Example 7.4: New Zealand skinks (cont.)

Using the data of Table 7.5, the numbers caught and the numbers of recaptures within the five days of the Otago trial are summarized in Table 7.6. The estimated number of skinks in the region investigated is

$$\widehat{N_S} = \frac{(188 \times 107) + (120 \times 277) + (76 \times 359) + (88 \times 390)}{1 + 18 + 38 + 45 + 52} = \frac{114960}{1 + 153} \approx 746.$$

Since $S = \sqrt{153/114960}$ the approximate 95% bounds are 643 and 889. The upper bound falls a good way short of the number estimated using model M_0 and suggests that nearly 60% of the skinks were caught at least once.

Table 7.6 The numbers of marked and unmarked skinks captured, or recaptured, on five successive days in a fenced plot in Otago, New Zealand.

Day, i	1	2	3	4	5
Number of marked skinks captured, m_i	0	18	38	45	52
Number of unmarked skinks captured, u_i	107	170	82	31	36
Total number of skinks captured, n_i	107	188	120	76	88
Number of marked individuals in population at day i, M_i	0	107	277	359	390

Model M_t assumes that all members of the population are equally likely to be captured. This implies that, as the number of marked individuals increases, so the proportion of marked individuals in future captures should increase. In this case the numbers of marked individuals at the times of the successive samples were 0, 107, 277, 359, and 390. The observed proportions of marked individuals, m_i/n_i, were 0, 18/188, 38/120, 45/76, and 52/88. The resulting plot, shown in Figure 7.2, suggests that the underlying assumption is reasonable, since the observed proportions are reasonably close to those expected (shown by the dotted line).

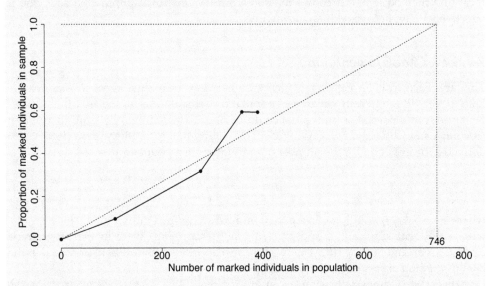

Figure 7.2 Plot of the proportion of marked individuals in the sample against the number of marked individuals in the population. The population size is estimated at 746. When all are marked, the proportion in a sample would be 1.0.

7.1.6 M_h models (Heterogeneous individuals)

Model M_t assumed that, although captures were more likely at some times than at others, at each time point they were the same for all individuals in the population. By contrast, models of the type M_h assume that, although the probability of capture is the same for an individual on every capture occasion, that probability varies from individual to individual.

In order to fit a model of the M_h type, it is necessary to make assumptions concerning the manner in which the capture probability varies across the individuals in the population under investigation. The possible approaches include:

1. Assume some particular type of probability distribution for the capture probabilities. Estimate any parameters of that distribution by using functions of $f_1, f_2, ..., f_t$, where f_i is the number of individuals captured on exactly i of the t trapping occasions. Hence estimate the population size.
2. Assume that the population consists of homogeneous groups of individuals with each group having its own capture probability. Estimate the number of groups and the capture probability for each group. Hence estimate the population size.

3. Assume that the capture probabilities are related to known characteristics of the individuals (e.g. sex). Estimate the parameters of the relation based on the characteristics of captured individuals. Estimate the underlying distribution of the characteristics in the population. Hence estimate the population size.

Different assumptions may lead to very different population estimates. There is no foolproof method. It will be wise to consider the results of several methods, but, unfortunately, the fact that two methods give similar results cannot be taken as evidence that their common estimate is correct. With all the models associated with the capture-recapture method, when it comes to estimating the unknown population size, one is effectively doing little more than gazing into a crystal ball!

7.1.6.1 Chao's lower bound

This approach to heterogeneity, which assumes that individual capture histories are known, provides an easily calculated population estimate. Assuming that the number of captures of a particular individual could be approximated by a Poisson distribution (Section 1.4.2) with the mean varying from one individual to another, Chao (1987, 1989) derived a lower bound for the number of individuals in a population as

$$\widehat{N_{LB}} = F + \frac{(t-1)f_1^2}{2tf_2} \tag{7.14}$$

where $F = \sum_{i=1}^{t} f_i = \sum_{i=1}^{t} u_i$ is the total number of trapped individuals. This formula, with a different application, appears with the description Chao2 in the programme SPADE (Chao and Shen, 2010). The formula also appears, with different notation, in the later discussion of species richness in Chapter 9.

Chao (1987) suggested using as an approximate 95% confidence interval for the lower bound

$$\left(F + \frac{f_1^2}{2f_2C}, \quad F + \frac{f_1^2C}{2f_2}\right), \tag{7.15}$$

where

$$C = \exp\left\{1.96\sqrt{\ln(1+\delta)}\right\},$$

and, following Böhning (2008),

$$\delta = \frac{1}{f_2} + \frac{4}{f_1} + \frac{2f_2}{f_1^2} - \frac{1}{F} - \frac{2}{2Ff_2 + f_1^2}.$$

A confidence interval for a lower bound is a rather tricky concept. The width of the interval is probably best regarded as an indication of the uncertainty inherent in the estimation process, though the lower side of the interval might be interpreted as truly being a lower bound on the number present.

Chao (2005) suggested a slight adjustment to $\widehat{N_{LB}}$ that results in a lowering of the lower bound. This is:

$$\widehat{N_{LB1}} = F + \frac{(t-1)f_1(f_1-1)}{2t(f_2+1)}. \tag{7.16}$$

If t is large so that $(t-1)/t \approx 1$, then Equations (7.14) and (7.16) simplify to give

$$\widehat{N_{LB2}} = F + \frac{f_1^2}{2f_2}, \tag{7.17}$$

$$\widehat{N_{LB3}} = F + \frac{f_1(f_1-1)}{2(f_2+1)}. \tag{7.18}$$

In her 1987 paper, Chao used $\widehat{N_{LB2}}$. In the programme *SPADE* (Chao and Shen, 2010), the authors use the description Chao1 for $\widehat{N_{LB2}}$ with the proviso that $2f_2$ is replaced by 2 if $f_2 = 0$. The *SPADE* programme also provides several bias-corrected versions of the estimators.

The use of any computer programme should note which of the alternative formulae is being used. Similar remarks apply to any confidence bounds.

Example 7.5: New Zealand skinks (cont.)

Using the data of Table 7.5, the numbers caught on exactly 1, 2, ..., 5 occasions, are summarized in Table 7.7. Using $\widehat{N_{LB}}$, the lower bound on the number of skinks in the fenced region is

$$426 + \frac{4 \times 332^2}{2 \times 5 \times 58} \approx 1186.$$

Using $\widehat{N_{LB1}}$, the estimate reduces to 1171. Using Equation (7.15), the approximate 95% confidence interval for the lower bound is (1112, 1742) This suggests that there were at least 1100 individuals present, despite only observing just over 400. It is also much higher than the estimates obtained using models M_0 and M_t.

Table 7.7 The numbers of skinks captured on exactly *i* of the five trapping days for the fenced plot in Otago, New Zealand.

i	1	2	3	4	5	Total (*F*)
Number trapped on exactly *i* occasions, f_i	332	58	17	15	4	426

7.1.6.2 The effect of varying the distribution for capture probabilities

Using simulations, Rivest and Baillargeon (2007) examined the consequences of varying the capture probability distribution. In their introduction, they observed that two models fitting the data equally well, may give very different estimates of population size; they concluded by stating: 'From a management point of view, a lower bound is much better than a larger speculative estimate whose validity is questionable.' The example that follows bears out these observations. Note, however, that any assumption of an underlying distribution should be justified.

Example 7.6: New Zealand skinks (cont.)

The programme *Rcapture* (Baillargeon and Rivest, 2007) can fit a variety of models with a single command. Table 7.8 presents an extract from the output, when the programme's closedp.t command is used with the skink data. For the model M_h, the programme offers four alternatives, with the first being Chao's lower bound. Details of the others are given in Rivest and Baillargeon (2007) and Baillargeon and Rivest (2007).

The estimates attributed to the M_0 and M_t models, differ from those obtained previously, because the programme uses a different estimation procedure. To decide which model best describes the data, it is customary to compare the AIC values. The AIC values for these two models are much greater than those for the M_h variants, suggesting that M_h is the better choice. The higher AIC value for M_t is largely due to its being 'expensive' in terms of the number of parameters used.

The best description (lowest AIC value) is provided by Chao's lower bound, with the estimate of 1186 obtained previously. A very close second (in terms of AIC values) is provided by the 'M_h Gamma3.5' model. However, this gives the implausible estimate that more than 7000 skinks were present. The estimation process hints at the implausiblity, since the estimate has an enormous standard error. Using the confidence interval based on 3 standard errors, the population size estimated by this model is roughly from 460 (implying nearly all were trapped at least once) to 15,000. By comparison, the interval based on Chao's lower bound is roughly from 800 to 1600, which is much more plausible.

The significance of the difference between the deviances of nested models, can be judged by comparison of that difference, with the upper percentage points of a chi-squared distribution. The improvement as a result of allowing the common capture probability to vary with time is $235 - 161 = 74$. The probability of obtaining that, or a greater value, from a chi-squared distribution with $29 - 25 = 4$ degrees of freedom, is of the order of 10^{-15}. The improvement using M_h with Chao's lower bound is yet more dramatic, since the probability of a chi-squared distribution with 2 degrees of freedom giving rise to a value exceeding 95 is less than 10^{-99}.

Table 7.8 Extract from the output of the *Rcapture* programme showing a range of estimates of the total population size, together with related statistics.

Model	Estimated abundance, $\hat{N}$	SE($\hat{N}$)	Deviance	df	AIC
M_0	752	44	235	29	336.2
M_t	733	42	161	25	270.2
M_h Chao (LB)	1186	133	140	27	245.6
M_h Poisson2	1169	116	159	28	261.8
M_h Darroch	2791	551	144	28	247.6
M_h Gamma3.5	7706	2416	143	28	245.8

The terms SE (standard error), deviance, df (degrees of freedom), and AIC were briefly introduced in Chapter 1.

Regardless of the accuracy of the estimates of population size, what these results do tell us, is that the variation between individual skinks is probably more relevant than the variation over time, though a model including both effects might be preferable to either.

7.1.7 Using M_h with explanatory variables

When individuals are captured it is usual to record characteristics such as gender, weight, probable age, and so forth. Some of these characteristics may be related to the probability of capture. When using R the relative importance of explanatory variables can be examined using the vglm command from the VGAM library. Details are given by Yee, Stoklosa, and Huggins (2015). This approach is surely preferable to the assumption of some distribution for the capture probabilities, if the choice of distribution is based on no clear reasoning.

Example 7.7: New Zealand skinks (cont.)

Among the variables recorded for the capture shinks were H (Habitat: in a gully, or on a ridge), S (Species: *Oligosoma maccanni*, *O. aff. polychroma* clade 5, or *O. inconspicuum*), and Z (Size: adult, or juvenile). With 426 skinks and 5 time points the aim is to use information on the variables H (1 d.f.), S (2 d.f.), and Z (1 d.f.) to give estimates of the 2130 values that record whether a skink is captured (1) or not (0). Denoting, for example, the model including both species and size by (S, Z), the AIC values and estimated population sizes, $\hat{N}$, for the eight possible models are given in Table 7.9.

The AIC values here are not comparable with those given in Table 7.8 because the underlying methodology is very different. The models having much the smallest AIC values are those that take note of the size (Z) of the skinks. A much smaller improvement results from including habitat information. Figure 7.3 illustrates the connections between the alternative models, with the line widths reflecting the differences in the AIC values for alternative models.

It is the juvenile skinks that were less likely to be caught more than once: of the juveniles, only 11% were caught more than once. For the adults, the corresponding proportion was 28%.

Note that the difference between the AIC values of a neighbouring pair of models, depends on which other variables are being considered. Thus, for two pairs that differ only by the inclusion of H, the difference for the models (Z) and (H, Z) is 6.2, whereas the difference between the models (S, Z) and (H, S, Z) is 1.3.

Table 7.9 The AIC values and population estimates for eight models that use information on habitat (H), species (S), and size (Z).

	–	H	S	Z	H, S	H, Z	S, Z	H, S, Z
AIC	2203.2	2197.4	2200.7	2158.2	2197.1	2152.0	2152.9	2151.6
$\hat{N}$	751	776	786	1025	1802	1067	1096	**1109**
SE($\hat{N}$)	43	50	58	135	62	147	161	**164**

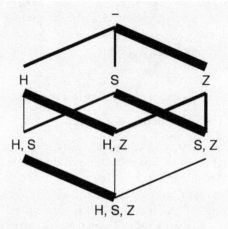

Figure 7.3 Line widths reflect the differences in the AIC values for alternative models. The prime importance of Z (size: adult or juvenile) is apparent.

7.1.8 Chao's lower bound for the model M_{th}

Chao, Lee, and Jeng (1992) suggested using

$$\widehat{N_{th}} = \frac{F + \gamma^2 f_1}{K} \tag{7.19}$$

where, as previously, f_i is the number of individuals trapped on exactly i occasions, and $F = \sum f_i$. The quantities K and γ^2 are given by

$$K = 1 - \frac{1}{T}\left(f_1 - \frac{2f_2}{t-1}\right), \tag{7.20}$$

$$\gamma^2 = \max\left(0, \frac{F}{K}\frac{\sum i(i-1)f_i}{(T^2 - \sum n_i^2)} - 1\right), \tag{7.21}$$

with n_i denoting the number of individuals trapped on the ith occasion, $T = \sum n_i$, and t being the number of trapping occasions. All summations are from $i = 1$ to $i = t$.

Example 7.8: New Zealand skinks (cont.)

The values of $n_1, \ldots, n_5$ were given in Table 7.6 and the values for $f_1, \ldots, f_5$ were given in Table 7.7. Using these values gives $K = 0.4767$ and $\gamma^2 = 0.6397$. Thus, using Equation (7.19), $\widehat{N_{th}} = (426 + 0.6397 \times 332)/0.4767 \approx 1339$. This is about 200 more than the estimates obtained previously.

7.1.9 Model selection

While the tree of models illustrated in Figure 7.1 summarizes the principal model types, it understates the complexity of the model selection process since, as previously noted, there can be many varieties of model M_h, and the same is true for models M_{th}, M_{bh} and M_{bth}.

Examination of the AIC values is useful, since this gives an idea of which factors are important for explaining the observed capture sequences. However, it requires a leap of faith, to assume that any capture-recapture model with parameters estimated from observed individuals, will apply equally to individuals so far unobserved.

As a final caveat, it should be noted that Grimm, Gruber, and Henle (2014) compared a dozen methods, using recapture data from populations of known size, and concluded that 'There is no single estimator that performs best and results in very good estimates for all data sets.'

7.1.10   Estimation based on individual encounters

In the previous sections it has been implicit that, at each trapping occasion, there will be many individuals trapped, with some being new individuals, and some being individuals previously caught. The skink data exemplified this situation. When physical contact may be undesirable for either the observed or the observer, 'capture' might take the form of remote observation, possibly with photographs, so that the 'captured' individuals may be unaware of their capture.

With large beasts, multiple captures may be unlikely: A. A. Milne's Pooh and Piglet did not expect to find several heffalumps in their heffalump trap. Instead, individuals may be identified singly. Methods include camera 'traps' (where movement triggers a picture of an individual identified by some characteristic pattern), and identification via DNA analysis (e.g. of faeces, hairs, or territory scent markings).

The special case where each capture refers to a single individual, was studied by Wilson and Collins (1992), who, using simulated populations, considered the bias and variability of a dozen different estimators. In this case, $n_i = 1$ for all i, and $t = T$. They found that the best-performing estimators were $\widehat{N_{DR}}$ given by Equation (7.10), and $\widehat{N_{LB2}}$ given by Equation (7.18). However, Keating et al. (2002) compared seven estimators and suggested that a more reliable estimator is $\widehat{N_{th}}$ given by Equation (7.19). This is unsurprising since it takes account of time variation and individual heterogeneity.

Example 7.9: Yellowstone grizzly bears

Keating et al. (2002) present data concerning observations of female grizzly bears that were accompanied by cubs in the Yellowstone National Park. All observations were made either in a designated grizzly bear recovery area, or a surrounding 10-mile buffer zone. Their data for 2001 are summarized in Table 7.10.

Using Equation (7.19), and calculating $K = 1 - (16 - 24/83)/84 = 0.817$, and $\gamma^2 = 0.335$, gives $\widehat{N_{LB2}} \approx 55$. This suggests that only about three-quarters of the bears present were detected.

Table 7.10 Summary of observations made in 2001 of the numbers of female grizzly bears with cubs observed in (or near) the Yellowstone Park grizzly bear recovery area. The quantity f_j is the number of bears sighted on exactly j occasions.

f_1	f_2	f_3	f_4	f_5	f_6	f_7	f_8	f_9	$F = \Sigma f_j$	$T = \Sigma j f_j$
16	12	8	0	1	1	0	0	1	39	84

7.2 Capture-recapture models for an open population

The methods of the previous section assumed that the population was unchanged from the first capture time to the last. In this section, that restriction is relaxed. The models allow for the possibility that, between the sampling times, some individuals in the original population may die, and some may move out of the study region. Equally, some new individuals may be born, or may enter the study region after the initial sample has been conducted. There may also be instances where an individual in the original population, leaves the study region, and subsequently returns.

7.2.1 The Jolly-Seber model

This model for an open population was proposed independently by Jolly (1965) and Seber (1965). Essentially the same model had been proposed earlier by Cormack (1964) in the context of estimating the probability of survival between successive captures.[1] The model is therefore also referred to as the *Cormack-Jolly-Seber model* or, more succinctly, the *CJS model*.

The key assumptions of the model are that, at each capture time, every member of the population then present:

- has the same probability, p, of being captured, and
- has the same probability, ϕ, of being in the population at the next capture time.

If p and ϕ are constant over time, then this is the equivalent of model M_0. If p varies from one capture period to the next, then this the equivalent of model M_t. If p varies with the type of individual, then this is a model of the M_h type. The equivalents of other closed-population models also occur.

An extra complication for the open-population models is that ϕ may vary from one capture period to another. This is likely to be true if the time intervals between successive captures are unequal, and may also occur if a capture site has experienced some major change (such as a fire or a flood) between successive capture times.

Denote the number of individuals in the ith sample by n_i, and assume that the population is sampled on $S(\geq 2)$ occasions. For a closed population, the population size at time t_i would be estimated by $n_i M_i / m_i$, with m_i being the number of marked individuals in the sample, and M_i being the number of marked individuals in the population at that time. However, for an open population, M_i is unknown, since some marked individuals may have left the population. Seber (1982) suggested estimating M_i using

$$\widehat{M_i} = \frac{(n_i - a_i) + 1}{R_i + 1} Z_i + m_i, \tag{7.22}$$

where

R_i is the number of individuals, captured in the ith sample, that were subsequently recaptured at least once ($R_S = 0$),

a_i is the number of accidental deaths or removals from the population that result from the ith capture, and

Z_i is the number of individuals marked *before* sample i, not captured in sample i, but recaptured subsequently ($Z_S = 0$).

In many cases $a_i = 0$. The population size, N_i, immediately prior to the ith sample ($i > 1$), is then estimated (Seber, 1982) by:

$$\widehat{N}_i = \frac{n_i + 1}{m_i + 1}\widehat{M}_i. \tag{7.23}$$

The approximate 95% confidence interval[2] for $\widehat{N}_i$, suggested by Manly (1984), has bounds

$$\frac{(4L_i + n_i)^2}{16L_i} \quad \text{and} \quad \frac{(4U_i + n_i)^2}{16U_i}, \tag{7.24}$$

where

$$L_i = \exp\left(T_i - 1.6S_i\right) \quad \text{and} \quad U_i = \exp\left(T_i + 2.4S_i\right). \tag{7.25}$$

Writing r_i for the number of those remaining in the population immediately after the ith capture (= $n_i - a_i$), the quantities S_i and T_i are given by

$$S_i = \sqrt{\frac{(\widehat{M}_i - m_i + r_i + 1)(r_i - R_i)}{(\widehat{M}_i + 1)(R_i + 1)(r_i + 1)} + \frac{n_i - m_i}{(n_i + 1)(m_i + 1)}}, \tag{7.26}$$

$$T_i = \log_e\left\{\frac{1}{2}\widehat{N}_i\left(1 - \frac{1}{2}\hat{p}_i + \sqrt{1 - \hat{p}_i}\right)\right\}, \tag{7.27}$$

where $\hat{p}_i$, the estimate of the proportion of the current population captured in the current sample, is given by:

$$\hat{p}_i = \min\left(1, \frac{n_i}{\widehat{N}_i}\right). \tag{7.28}$$

Example 7.10: Grizzly bears in Banff National Park

Whittington and Sawaya (2015) compared a variety of capture-recapture models using simulated data. They also used data, collected between 2006 and 2008, that concerned the grizzly bears in Banff National Park, Alberta. Table 7.11 summarizes that portion of their data that refers to the seven sampling occasions, between May and October of 2007, in which hair samples were collected from 'bear rubs'. This subset of the data refers to 46 individual bears whose identities were determined by DNA analyses of the collected hairs.

Table 7.11 Capture histories for grizzly bears in the Banff National Park, Alberta. The data result from the DNA analyses of hair samples collected from 'bear rubs' on seven sampling occasions during 2007. There were 46 bears identified.

1000000	2	0100000	2	0010000	5	1010000	1	0110000	2
1110000	1	0001000	2	1001000	1	0101000	2	0011000	3
0111000	1	1111000	2	0000100	3	0010100	1	1001100	1
0101100	1	0000010	2	1010010	1	0001010	2	1001010	1
0111010	1	0000110	1	1000110	1	0100110	2	1100110	1
1110110	1	0011110	1	0111110	1	1100001	1		

The first step is to use the capture histories to calculate the values of n_i, m_i, R_i, and Z_i. The results are given in the first part of Table 7.12. The second part of the table reports the results of subsequent calculations. While it is important to carry through calculations with maximum precision, a publicity report should reflect the uncertainty concerning the appropriateness of a model. No mathematical model can hope to perfectly capture the infinite variety of nature, and its evolution through space and time. It might, for example, be appropriate to report the number of bears present in period 5, as being between 30 and 70, with a best estimate of 40.

Table 7.12 Summary statistics for the grizzly bear capture histories in Table 7.11.

Sample	i	1	2	3	4	5	6	7
No. in sample	n_i	14	18	19	18	14	16	12
No. previously identified	m_i		6	11	13	10	14	11
No. also seen in later sample	R_i	12	16	12	12	8	5	
No. not in sample but seen both before and after	Z_i		6	11	10	12	6	
Estimated number of marked individuals at time of sample	$\hat{M}_i$		12.7	27.9	27.6	30	31.0	11.0
Estimated population size at time of sample	$\hat{N}_i$		34.5	46.6	37.5	40.9	35.1	11.9
Estimated propn. sampled	$\hat{p}_i$		0.52	0.41	0.48	0.34	0.46	1.00
Lower bound for $\hat{N}_i$			25.2	34.5	29.6	29.2	24.2	21.6
Upper bound for $\hat{N}_i$			62.4	77.8	56.2	71.7	69.7	39.4

7.2.2 Computer programmes for open-population mark-recapture analysis

The most used computer programme for mark-recapture estimation of population size for open populations is probably *MARK*, for which a manual (of greater than 1000 pages) is freely available from http://www.phidot.org/software/mark/docs/book/. A front end for R users is provided by the *RMark* package. Other R-based packages include *CARE1*, *mra*, *Rcapture*, and *secr*. All are freely downloadable from CRAN (The Comprehensive R Archive Network).

Example 7.11: Australian frogs

An investigation of the health and survival rates of the endangered Australian frog *Litoria raniformis* took place during 2004 and 2005 at sites within the Merri Creek catchment to the north of Melbourne, Victoria, Australia. Table 7.13 gives the frog counts for site 6, a stream sampled on ten occasions in 2005. The complete data were analysed in detail by Heard et al. (2014).

Table 7.13 Numbers of frogs caught in a particular stream in the Merri Creek catchment on ten occasions in 2005. There were 45 frogs caught at least once. Of these, 11 were caught twice, 6 were caught on three occasions, and 2 were frogs caught four times.

Capture periods	1	2	3	4	5	6	7	8	9	10
Number of frogs captured	3	9	6	14	8	7	13	3	9	2

Using the R package *mra*, the analysis begins by fitting the M_0 model:

```
library(mra)

# seen is a 45 × 10 matrix of capture histories
# frogdays is a 45 × 9 matrix of the time intervals
# in days between successive capture sessions

F.cjs.estim(capture=~1,survival=~1, histories=seen,
intervals=frogdays)
```

The resulting output includes the following:

```
Capture var   Est       SE
(Intercept)   -0.91964  0.33146

Survival var  Est       SE
(Intercept)   2.36355   0.34312

Link = logit
Model df =  2
Log likelihood =  -79.9055293332119
AIC =  163.811058666424
```

As the output suggests, the model uses logits (Section 1.7). The estimates are $\log\{p/(1-p)\} = 0.91964$, and $\log\{\phi/(1-\phi)\} = 2.36355$, where here ϕ is the probability of survival from one day to the next. The standard errors (SE) are relatively large, suggesting that there is a good deal of uncertainty concerning the estimated values. Using these values the probability of capturing a frog that is present at the time of capturing is estimated as

$$\hat{p} = \exp(-0.91964)/\{1 + \exp(-0.91964)\} = 0.29,$$

and the probability of a frog surviving from one day to the next is estimated as

$$\hat{\phi} = \exp(2.36355)/\{1 + \exp(2.36355)\} = 0.91.$$

The model presumes that these estimates hold for all frogs throughout the survey period.

We next examine whether, for this site, there is evidence that the Age class of an individual (0 = juvenile, 1 = subadult/adult) affects survival or catchability. The AIC values for the four relevant models (which are therefore of the M_h type) are given in Table 7.14.

Table 7.14 AIC values for models investigating whether the age of a frog (juvenile or subadult/adult) affects the probabilities of capture (p) or survival (φ) or both.

Probability affected	Neither	Just capture	Just survival	Both
AIC value	163.8	163.4	164.3	165.3

The model that allows different values for p has the lowest AIC value. The code required to fit this model is:

```
model<-F.cjs.estim(capture=~ivar(age,10),survival=~1,
histories=seen,intervals=frogdays)
model

# age is the 45 × 1 vector of age identifiers which is
#connected to the 10 capture sessions by the ivar function
```

The output is now a little more complicated:

```
Capture var    Est       SE
(Intercept)    -0.79141   0.34509
ivar(age, 10)  -1.28165   0.86069

Survival var   Est       SE
(Intercept)    2.4035    0.35055

Link = logit
Model df =  3
Log likelihood =   -78.6778568595745
AIC =   163.355713719149
```

The model df has increased from 2 to 3 since there is an extra parameter that measures how the probability of capture for juveniles differs from that for adult frogs. The ivar(age, 10) line in the output gives information about how the second category of the variable concerned (age) differs from the first. If there were no difference, then the estimate would be near zero, and the ratio of the estimate to its standard error would be less than 1. Here that ratio is greater than 1 and this is reflected in the improved fit and reduced AIC value. The two capture probabilities are estimated as follows:

$$\hat{p}_{adult} = \exp(-0.79141)/\{1 + \exp(-0.79141)\} = 0.31,$$
$$\hat{p}_{juv.} = \exp(-0.79141 - 1.28165)/\{1 + \exp(-0.79141 - 1.28165)\} = 0.11.$$

These estimates reflect the fact that 35 of the 45 captured frogs were adults, with 17 of these being recaptured on at least one occasion. By contrast, there were only two recaptures of the 10 juveniles.

Using this model, the following R code gives the abundance estimates and the associated lower and upper 95% values:

```
model$n.hat
model$n.hat.lower
model$n.hat.upper
```

Figure 7.4 illustrates the results. Taking account of the uncertainty in the individual estimates, it appears that the population size was essentially constant and probably amounted to between 20 and 40 individuals.

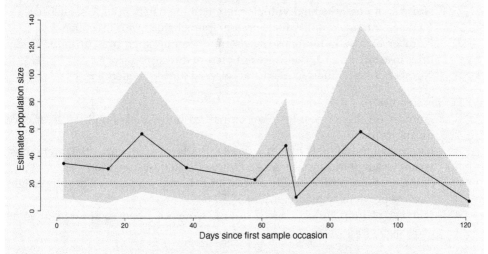

Figure 7.4 The estimated total numbers of frogs during the 2005 season using a model that assumes that adults and juveniles have different probabilities of selection. The shaded region indicates the 95% confidence intervals for the estimates.

7.3 Pollock's robust design

Whether a population is considered to be open or closed is dependent upon the timescale over which sampling takes place. If samples are taken on successive days, then it would seem reasonable to assume that the same individuals were present throughout. However, if a population is only sampled every six months, then one can expect births, deaths, immigration and emigration to have taken place.

Pollock (1982) proposed a sampling scheme that observed the population over both short and long timescales. At intervals well-separated in time, there would be short sequences of samples. Within each sequence, the population could be regarded as closed. Between sequences, the population would be regarded as being open. Pollock suggested using closed-population methods for dealing with each sequence, and the Jolly-Seber model for handling the changes in population between sequences. Pollock's approach is usually referred to as the *robust design*.

7.3.1 *The integrated design

Kendall, Pollock, and Brownie (1995) developed integrated models that take account of the entire capture history. Within a sequence, any of the models M_0, M_t, ... may be appropriate, but, over time, the parameters in these models may change: the authors list 24 model combinations that they felt might apply. There are therefore many aspects to consider (as the following makes clear).

- For the population:
  U_i Number of unmarked individuals present immediately prior to sequence i.
  M_i Number of marked individuals present immediately prior to sequence i.
  B_i Number of new individuals (newborn, or immigrants) present immediately prior to sequence $i+1$, that were not present during sequence i.
  R_i Number of individuals marked and released during sequence i.

- For the samples:
  u_i Number (in the range 0 to U_i) of unmarked individuals caught at least once during sequence i.
  m_{hi} Number of individuals caught at least once during sequence i that were most recently caught during sequence h (where h is in the range 1 to $i-1$).
  m_{+i} Number of individuals caught at least once during sequence i that were caught in some previous sequence: $m_{+i} = \sum_{h=1}^{i-1} m_{hi}$.
  r_i Number of individuals marked during sequence i and recaptured in some subsequent sequence.

- For the sample sequences:
  $x_{0i}(c)$ Number of individuals unmarked prior to sequence i that exhibit the capture history c during sequence i.
  $x_{hi}(c)$ Number of individuals most recently captured during sequence h that exhibit the capture history c during sequence i.
  f_{ij}^0 Number of individuals unmarked prior to sequence i that were captured during sample j of sequence i.
  f_{ij}^h Number of individuals most recently captured during sequence h that were captured during sample j of sequence i.
  l_i The number of samples in sequence i.
  k The number of sequences.

- For the probabilities within a sequence:
  p_{ij} The probability that an individual, present during sequence i, is captured for the first time in that sequence, in sample j.
  c_{ij} The probability that an individual, already captured during sequence i, is recaptured in sample j.
  p_i^* The probability that an individual is captured during sequence i: $p_i^* = 1 - (1 - p_{i1})(1 - p_{i2}) \cdots (1 - p_{il})$

- For the probabilities between sequences:
  ϕ_i The probability that an individual, in the population midway through sequence i, is also in the population midway through sequence $i+1$.
  χ_i The probability that an individual, alive at the end of sequence i, is never seen again.

The joint likelihood then has three components:

1. The product, from $i=1$ to $i=k$, of the probabilities of selecting u_i unmarked individuals from the U_i available ($i = 1, 2, ..., k$).
2. The product of the probabilities that the R_i ($i = 1, 2, ..., k - 1$) individuals initially captured in sequence i, are next recaptured during series j ($j = i + 1, i + 2, ..., k$).
3. The product, from $i = 1$ to $i = k$, of the probabilities of the various possible within-sequence recapture histories for the u_i individuals captured at least once during sequence i.

In practice some individuals may be unavailable for capture during some sequences as a consequence of temporary emigration. This scenario was addressed by Kendall, Nichols, and Hines (1997), and by Kendall and Bjorkland (2001), with a concomitant increase in necessary notation.

7.4 Spatial capture-recapture models

Using a closed-population model, the size of the region being sampled is usually well-defined. However, this is not the case for an open population with mobile individuals. For any given location within the sampling region, in addition to factors such as the age and gender of an individual, the probability of observing an individual, p, will be dependent on the distance, d, between the sampling location and the centre of that individual's home range. At a random time point an individual is more likely to be near the centre of its territory than some arbitrary location near the territory's edge. For this reason, the probability of detecting an individual at a sampling point is a function of d, being greatest when d is smallest.

Figure 7.5 illustrates two typical *detection functions*: the half-normal (proportional to $\exp(-ad^2)$) and the exponential (proportional to $\exp(-bd)$), where a and b are constants to be estimated from the data.

The use of spatial information in the capture-recapture context was first suggested by Efford (2004), and was developed independently by Royle and Young (2008), and Borchers and Efford (2008). Both pairs of authors used data from a spatial array of traps, to estimate the number of territories with centres lying inside the study region.

With multiple traps the data consists of triples of information: the identity of the captured individual, the time of capture, and the location of capture. When an individual is recaptured, the locations of its capture sites gives an indication of the mobility of that

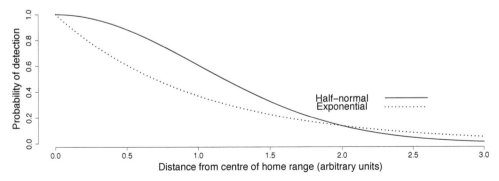

Figure 7.5 A comparison of the half-normal and exponential detection functions.

individual, and, thereby, of similar individuals. As well as physical traps that result in a creature's capture, the models can be used when the 'traps' consist of locations where a creature's DNA is found, or where remote cameras have been triggered by movement.

Royle and Young (2008) proposed a model that assumed static circular territories uniformly distributed over a large area that included the study region. At any given time, the displacement of a creature from the centre of the territory, was modelled using orthogonal normal distributions. Royle, Fuller, and Sutherland (2016) provide BUGS code[3] for this model, and also for the case where the activity centres move during the sampling period.

An alternative is provided for R programmers by the *secr* package introduced by Efford (2011). Building on the previous approach of Borchers and Efford (2008), this package fits models for search data where a polygonal region is searched repeatedly, or where a single search results in multiple DNA traces. It can also be used with data from linear transects, with binary data (animal detected or not detected on each occasion), or with count data (multiple detections per occasion). Efford (2019) suggests that the assumption of circular home ranges (implied by detection functions that depend only on the distance from a central point) does not greatly effect the accuracy of the estimates obtained.

Example 7.12: Grizzly bears in Banff National Park (cont.)

The previous analysis (Example 7.10) of the data on bear rubs ignored their locations by using only information, for each sampling period, on whether a bear was detected somewhere in the region. The results of that analysis, given in Table 7.12, showed how the numbers fluctuated during the season with a midsummer peak.

Figure 7.6 (a) shows the locations of all the rubs used by at least one bear during the summer of 2007, with those visited by bear 50 being emphasized. Figure 7.6 (b) focuses on bear 50 and indicates the average positions of the rubs visited on each of the seven sampling periods. The capture history data summarized in Table 7.11, showed that 16 of the 46 bears were identified on only one occasion during the year. The wanderings of bear 50 illustrate the general mobility of the population.

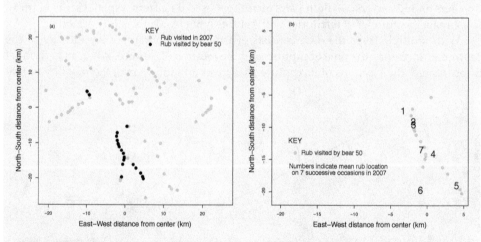

Figure 7.6 (a) The locations of bear rubs in Banff National Park, Alberta, Canada in the summer of 2007 with rubs visited by bear 50 emphasized. (b) The average position of the rubs visited by bear 50 during seven successive periods (scales differ).

The previous analysis estimated the numbers of bears present at different periods during the year. The estimates for the seven individual periods peaked at about 47, with an upper bound at 78. The overall estimate of the number of bears present at some time during the year may be expected to be higher. For this the *secr* programme will be used.

The *secr* programme requires the construction of two files. The trap file has three columns: trap identifier, east–west location, north–south location. The capture file has 4 columns: survey number, animal identifier, capture occasion, trap identifier. Here is an extract from this file showing some records for bear 50:

```
1 50 4 88
1 50 4 188
1 50 4 195
1 50 4 208
1 50 4 219
1 50 5 189
1 50 5 192
1 50 5 193
1 50 6 208
```

Bear 50 was detected just once in period 6. This was at location 208, which the bear also visited in period 4.

Denoting the two required files in an obvious way, the following R commands can be used:

```
library(secr)
beardata<-read.capthist("captfile","trapfile",
detector="proximity")
secr.fit(beardata, buffer=30000, detectfn=0)
```

Here the detector="proximity" statement informs the programme that an individual may be 'trapped' at several locations during a time period. The choice detectfn=0 instructs the programme to use a half-normal detection function.

The resulting lengthy output records that 46 animals were detected on 7 occasions with a total of 226 'sightings'. The output records many details of the model fitted including the following:

```
Mask area       :   855596.2 ha
AICc            :   2044.128
Fitted parameters evaluated at base levels of covariates
      link    estimate  SE.estimate          lcl          ucl
D     log 1.233009e-04 1.952054e-05 9.058165e-05 1.678387e-04
```

The programme has superimposed a 'mask' around the observed sightings. In this case the mask has an area of 855596.2 ha. The density of the bears has been estimated as '1.233009e-04' per hectare, so that the estimated number of bears in the region is

$$855596.2 \times 1.233009 \times 10^{-4} \approx 105 \text{ bears.}$$

The lower and upper 95% confidence limits for the density correspond to counts of 78 and 144 bears.

The extract of the output refers to 'covariates'. A *covariate* is a quantity that is of interest because its value may have an effect on the value of the variable of primary interest (which, in this case, is abundance). In this example there are no covariates, as the model fitted corresponds roughly to the model M_0 for a closed population. The *secr* programme allows for the parameters of the detection function to vary with, for example, the age or gender of an individual (the equivalent of model M_h). Age and gender would then be described as covariates. As previously, the choice between alternative models would be guided by a comparison of the AICc values.

7.5   Mark-resight estimation

If capturing and marking the individuals in a population is expensive, or is potentially disruptive to the population, then the mark-resight approach is appealing, since it requires only a single initial marking event. All subsequent sampling consists of recording the numbers of marked and unmarked individuals. The analysis depends on whether the marked individuals can be distinguished from one another (e.g. by using numbered tags) or are otherwise indistinguishable (e.g. by using coloured rings).

Although the practical aspects of the capture-recapture and mark-resight methods are distinctly different, the analyses are essentially identical, since the only difference is that the number of marked individuals is constant rather than steadily increasing. One consequence is that some formulae simplify. For example, the Chapman-corrected Schnabel formula given by Equation (7.11) becomes

$$\widehat{N_S} = \frac{nM}{1+m},\qquad(7.29)$$

where n is the total number of individuals sighted, m is the total number of marked individuals sighted, and M is the number initially marked. The approximate 95% confidence interval simplifies to

$$\left(\frac{nM}{1+m+2\sqrt{m}},\ \frac{nM}{1+m-2\sqrt{m}}\right).\qquad(7.30)$$

Example 7.13: New Zealand skinks (cont.)

The skink data consisted of counts on five successive days. The data, which were summarized in Table 7.6, are presented in mark-resight form in Table 7.15.

Using these data, the estimated number of skinks in the region investigated is $472 \times 107/59 = 856$, with approximate 95% confidence bounds as 680 and 1154. The method gives a much wider confidence interval than before (it was previously 643 to 889) because the information provided by marking the new individuals caught after the first day has been ignored.

Table 7.15 The skink capture data presented as mark-resight data (by ignoring markings other than the 107 made during the initial capture).

Resight periods	1	2	3	4	Total
Number of skinks captured	188	120	76	88	$n = 472$
Number of marked skinks captured	18	14	15	11	$m = 58$

7.5.1 An unknown number of marked individuals

Arnason, Schwarz, and Gerrard (1991) were the first to address the problem of estimating the size of a closed population by observing the numbers of marked individuals, when the total number of marked individuals in the population was unknown. They assumed that every individual had the same probability of being sighted. McClintock et al. (2009) addressed the limitations of the earlier approach within the framework of the robust model (Section 7.3). McClintock, White, and Pryde (2019) presented an improved model involving a Poisson-lognormal distribution (Section 1.5.2) with an inflated proportion of zero sightings.

8. Distance methods

Capture-recapture methods can only be used if individuals are distinguishable from one another. If capturing implies physical trapping, then this might harm the individual concerned and could have wider implications for the population under investigation. Distance methods provide a means of avoiding these difficulties, though they are not without their own problems. As the name suggests, distance methods require a distance to be assigned for each individual sighted. The distance must be numeric, but need not be precise: a distance range will suffice.

With distance methods there are no simple formulae. Instead, it is the computer that does the work. There are several freely available computer programmes available to do the calculations. They all use similar methods. The best documented, with a considerable online presence of worked examples and video tutorials, is the programme *Distance*. Most of the examples in this chapter use that programme. There is a Windows version available for those who prefer it.

8.1 The underlying idea

Suppose that 23 individuals are observed at a distance of 100 m from an observer. If it is known that only a quarter of the individuals at that distance will be observed, then the best guess of the number of individuals actually present at that distance will be $23 \times 4 = 92$ individuals.

In general, if $n(d)$ individuals are observed at a distance d and $p(d)$ is the probability that an individual at that distance will be observed, then the best estimate of the number present at that distance will be

$$n(d)/p(d).$$

If m individuals are observed, with individual i being at a distance d_i, then the total number of individuals will be estimated by

$$\sum_{i=1}^{m} d_i/p(d_i).$$

The function $p(d)$ is called the *detection function*. The focus of distance sampling is the selection of a detection function that provides a credible fit to the observed data.

8.2 Detection functions

Two general types of detection function (the exponential and half-normal) were illustrated in Figure 7.5, though the exponential function is rarely appropriate. Two other possibilities are the uniform detection function which assumes that all individuals are detected up to some critical distance, with none being detected at greater distances, and the two-parameter hazard-rate function

$$1 - \exp(-(d/a)^{-b}),$$

which has a flexible shape determined by the values chosen for the two parameters a and b as illustrated in Figure 8.1.

The fit of any of these functions may be improved by the addition of extra terms. Using the terminology of Buckland et al. (2001), the uniform, half-normal, and hazard-rate functions are described as *key functions*, with the complete detection function being given by the product:

$$\text{Detection function} \propto \text{Key function} \times (1 + \text{Series expansion}). \qquad (8.1)$$

Two examples of the *series expansions* used are

$$\text{Cosine:} \quad a_1 \cos\left(\tfrac{\pi d}{w}\right) + a_2 \cos\left(\tfrac{2\pi d}{w}\right) + \cdots, \qquad (8.2)$$

$$\text{Polynomial:} \quad a_1 \left(\tfrac{d}{w}\right)^2 + a_2 \left(\tfrac{d}{w}\right)^4 + \cdots, \qquad (8.3)$$

where $a_1, a_2, \ldots$, are constants, and w is the maximum distance at which detections may occur. A third alternative is provided by Hermite polynomials.[1]

In practice, the computer routines that exist for this methodology have these key functions and series expansions internally programmed, so that the user can concentrate on the results rather than the mathematical mechanics. In this chapter most analyses will use the *Distance* programme because of the very large amount of helpful online material. Alternatives include the *unmarked* and *Rdistance* programmes which have similar capabilities.

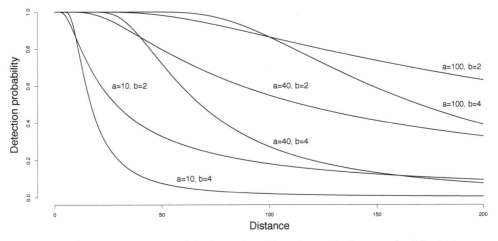

Figure 8.1 The two parameters of the hazard-rate function make it a very flexible choice as a detection function.

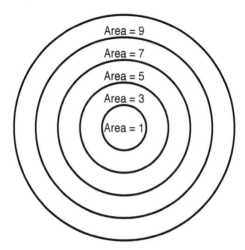

Figure 8.2 If individuals were uniformly distributed, and every individual were detected, then there would be increasing numbers observed with increasing distance from the central point.

8.3   Point transects

Figure 8.2 shows a series of rings around the central sampling point. If a ring has inner and outer boundaries at distances d and $d+1$, then the area of the ring is

$$\pi(d + 1)^2 - \pi d^2 = \pi(2d + 1),$$

so that each ring has a larger area than its predecessor. If individuals were randomly scattered, then (apart from random variation) this would mean that the numbers of individuals would increase with increasing distance from the sampling point. However, because the probability of detection reduces with distance from the sampling point, the numbers of individuals detected, as distance increases, typically starts low, rises to a peak, and then falls towards zero.

At short distances most individuals present will be observed, whereas, at long distances, it will be literally a matter of chance as to which of the individuals there present are actually detected. For that reason, Buckland et al. (2001) suggest ignoring the largest 5% to 10% of distances.[2]

Example 8.1: Japanese white-eyes in Hawai'i

Japanese white-eyes (*Zosterops japonicus*) are common in Hawai'i and may provide a useful indicator of changes in bird populations as a consequence of habitat change. The data here are related to the study by Camp et al. (2019) concerning the effect of tree mortality on the local bird population. Data were collected in 2003 and 2016. Japanese white-eyes were observed at 220 sites of the 223 sites visited in 2016. Figure 8.3 presents a histogram of the 2016 distance data.

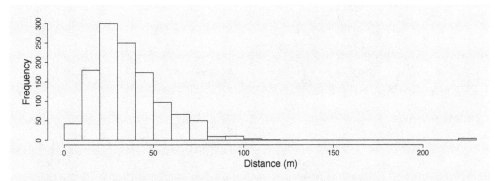

Figure 8.3 Histogram of the distances at which 1186 Japanese white-eyes were observed during the 2016 survey in Hawai'i.

8.3.1 Using *Distance* for point transects

The *Distance* programme was originally developed for use with Windows, but it is now available for users of R. The originators have produced a huge amount of online material including slides, videos, and worked examples. For information the reader should start at the website http://distancesampling.org. What follows here is just an indicator of the possibilities, and the general approach to modelling distance data; it is not a substitute for the detailed online advice.[3]

The usual steps are as follows:

1. Plot the distance data (e.g. via the hist command) in order to gain an understanding of its shape, and to identify any outliers.
2. Fit a detection function so as to assess the appropriate amount of data truncation.
3. Fit alternative models and choose one or more candidates on the basis of the AIC values (see Section 1.9).
4. Examine the chosen model(s) to determine the estimated density of individuals in the vicinity of the sampling points.

Example 8.2: Japanese white-eyes in Hawai'i (cont.)

Figure 8.3 showed that most birds were seen at distances around 20 m, with only a few at distances greater than 80 m. There are also a few suspiciously extreme values. The R code begins by overwriting information about other species, while preserving information about the sampling points, and then placing it in a dataframe:

```
Data for many species are in an array called data2016
having n rows. The next two lines overwrite the names and
distances for all except the chosen species, JAWE

w<-which(data2016[,6]!='JAWE')
data2016[w,6]<-NA; data2016[w,5]<-NA
```

Create required vectors

```
Sample.Label<-data2016[,3]; distance<-jdata2016[,5]
Area<-rep(1,times=n); Effort<-rep(1,times=n)
Region.Label<-rep('Hawaii',times=n)
```

Store in dataframe

```
data2016DF<-data.frame(Sample.Label, distance, Area,
Effort, Region.Label)
```

The five vectors Sample.Label, distance, Region.Label, Area, and Effort are required inputs. Area is either set equal to the area of the region being sampled, or (as here) to the value 1. When set equal to 1 the output refers to density rather than abundance. The vector Effort indicates the numbers of times that each sampling point was visited. In this case each point was visited once.

To investigate the likely fit of a detection function to the data, an initial truncation at 80 was used within the ds() command and a histogram of the data together with the fitted detection function was produced. This showed that a more extreme truncation was required, so that the exercise was repeated with a 10% truncation using the following R code:

```
library("Distance")

u1c10 <-ds(data2016DF, transect="point", key="unif",
adjustment="cos", order=1, truncation="10%")

plot(u1c10); points(c(0,100), c(0.1,0.1), type='l')

summary(u1c10)
```

The resulting diagram is shown as Figure 8.4. The output from the summary command includes the following:

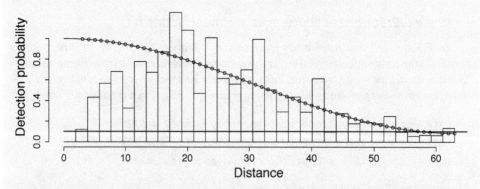

Figure 8.4 The fit of the detection function u1c10 consisting of the uniform key plus a single cosine term, with the 10% most extreme values ignored. Also shown is the line corresponding to p(*d*) = 0.1.

```
Number of observations :  1078
Distance range         :  0 - 63

Model : Uniform key function with cosine adjustment
           term of order 1

AIC   : 8584.888

Adjustment term coefficient(s):
            estimate          se
cos, order 1 0.8440258 0.01908081
```

Using this output, with d denoting distance, the dotted line in Figure 8.4 is given by

$$\{1 + 0.8440 \cos(d\pi/63)\}/1.8440.$$

A measure of the fit of alternative models is provided by the AIC value, with smaller values preferred. The outputs from the four models with the lowest AIC values are summarized in Table 8.1. Each model summary gives values for 'N in covered region' and 'Average p', which appear as $\hat{N}$ and $\hat{P}$ in the table. The quantity $\hat{P}$ requires explanation: it is the estimated probability that an individual, present in the covered region, will be detected. The estimated number, $\hat{N}$, is simply the observed number divided by $\hat{P}$.

Table 8.1 shows that although the models give reassuringly similar values for $\hat{N}$, the estimate provided by hr10 has a much smaller standard error. Figure 8.5 illustrates the corresponding probability density functions.[4]

Rather than simply eyeballing the fit, one can use the following command to examine the q-q plot (Section 1.10):

```
gof ds(hr10)
```

This produces Figure 8.6 and gives the output:

```
Distance sampling Cramér-von Mises test (unweighted)
Test statistic = 0.38878 p-value = 0.0773642
```

Table 8.1 The four best models for the Japanese white-eye data (after 10% truncation).

Model	Key	Type	No. of terms	AIC	$\hat{N}$	s.e.	$\hat{P}$	s.e.
hr10	Hazard rate	None		8566	2397	111	0.45	0.018
u4c10	Uniform	Cos	4	8571	2400	408	0.45	0.075
u3p10	Uniform	Poly	3	8579	2678	180	0.40	0.025
hn2c10	Half-normal	Cos	2	8579	2583	288	0.42	0.046

Column group headers: Adjustment (Type, No. of terms); Estimated no. in surveyed area ($\hat{N}$, s.e.); Estd. detect. prob. in surveyed area ($\hat{P}$, s.e.)

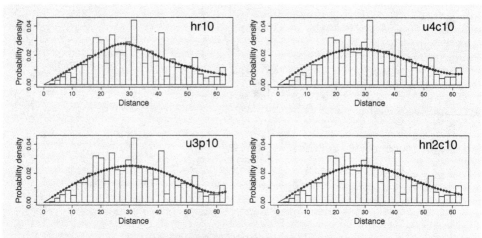

Figure 8.5 The fits of four alternative probability density functions to the distance data (after truncation at the 10% level). The best fit is provided by hr10.

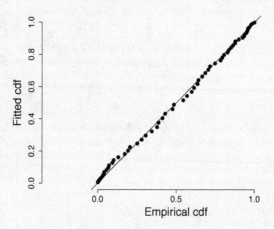

Figure 8.6 The q-q plot resulting from the use of hr10 as the detection function for the white-eye data.

This q-q plot shows that the points lie very close to the line of equality. The value of the Cramér-von Mises statistic is small (by chance, when the model fits, the output states that this value would be exceeded on only about 8% of occasions) suggesting that the model is an acceptable fit.

The summary output stated that, after truncation, the distance range was from 0 m to 63 m. Since there were 223 sampling points, this implies that the total area surveyed was $223 \times \pi 63^2$ m², which is about 278 hectares. With $\hat{N} = 2397$, this implies that, in 2016, the density of Japanese white-eyes was about 8.6 per hectare.[5] More accurate values (but given with far too many decimal places) are provided by the summary(hr10) command:

```
Summary statistics:
 Region Area CoveredArea Effort    n   k
1 Hawaii   1   278.0583 223 1078 223
```

```
Density:
   Label Estimate       se         cv       lcl       ucl
1 Total 8.620282 0.4497381 0.05217208 7.781912 9.548973
```

The approximate 95% confidence interval is a density of between 7.8 and 9.5 birds per hectare.

8.4   Using imprecise distance data

It will not always be possible to record distance data with great precision. Indeed, for 'citizen science' projects, such as the UK's breeding Bird Survey, the volunteer observers are only required to use three distance classes: nearer than 25 m, between 25 and 100 m, and further than 100 m. Despite the apparent vagueness of these classes, useful estimates can still be obtained, as the following example demonstrates.

Example 8.3: Japanese white-eyes in Hawai'i (cont.)

The white-eye data were recorded to the nearest metre. Suppose, instead, that distances had been recorded using five-metre bins: $d < 5$ m, $5 \le d < 10$ m, etc. Omitting four outlier records, the frequencies for the 2016 sample are given in Table 8.2.

The *Distance* programme uses the same commands as previously, together with information concerning the groups used. The lowest AIC value resulted from using the hazard-rate key function with cosine extensions. The essential code is

```
binends<-c(0:5)*24

hrcbin<-ds(jaweDF, transect="point", key="hr", adjustment="cos",
cutpoints=binends)
```

The summary took the usual form, reporting an estimate of 8.0 birds per ha, with an approximate 95% confidence interval of 7.3 to 8.7 birds per ha. This compares with the estimate for the ungrouped data, which was (7.7, 9.5).

To examine the influence of the choice of distance groups, the 24 groups used in the preceding analysis were replaced by the three groups used by the British Breeding Bird Survey. Following the same procedure, the approximate 95% confidence interval is astonishingly similar: (7.6, 9.4).

Table 8.2 The 2016 Japanese white-eye distance data summarized using five-metre bins.

Bin top	5	10	15	20	25	30	35	40	45	50	55	60
Frequency	3	32	43	130	163	142	149	98	109	75	67	32
Bin top	65	70	75	80	85	90	95	100	105	110	115	120
Frequency	37	29	33	17	4	5	7	3	1	1	1	1

8.5 Introducing covariates

The ease with which individuals are detected will depend on factors such as the skill of the observer, the time of day, the type of location, and the weather. Multiplicative factors corresponding to any, or all, of these covariates can be easily introduced into the analysis, which might proceed as follows:

1. Examine the fit of any of the detection functions in order to determine the appropriate truncation level.
2. Determine which of the half-normal and hazard-rate keys performs best when used with no extensions.[6]
3. Refit the best model with the covariate included.
4. If the AIC value for the key function with no covariates is the lower, then the covariates need not be taken into account.

Example 8.4: Nightingale reed warblers in Alamagan

Alamagan, one of the Northern Mariana Islands, is one of the few locations where the nightingale reed warbler (*Acrocephalus luscinia*) can be found. Data from a survey conducted in 2010 can be retrieved from http://www.sciencebase.gov. There were two groups of surveyors. One group were from the Division of Fish and Wildlife (DFW) of the Commonwealth of the Northern Mariana Islands, and the other were from the US Fish and Wildlife Service (FWS). Their skill levels might differ, so the analysis now treats the groups of surveyors as a covariate.

A total of 153 detections were made at 121 sampling points, though there were no records for 60 of those points.[7] Figure 8.7 suggests that the analysis will find little difference in the distributions of the distances recorded by the two groups of surveyors.

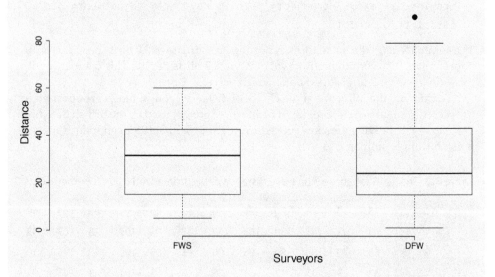

Figure 8.7 The boxplot comparing the distances recorded by the two sets of observers.

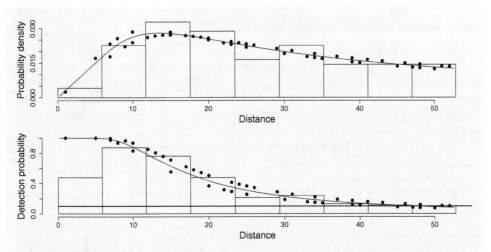

Figure 8.8 The density and detection functions using the hr10G model. On each diagram, the two sets of dots indicate the fitted values for the two groups of observers. There is little difference between them.

Information on the survey teams was held in the vector Group which formed part of the dataframe dataDF. A preliminary investigation suggested that, as before, a 10% truncation was suitable. To introduce the covariate, the relevant R command is:

```
hr10G<-ds(dataDF, key= 'hr', transect="point",
adjustment=NULL, truncation="10%", formula=~Group)
```

The fitted density and detection functions are illustrated in Figure 8.8. Since the AIC value for model hr10G is greater than that for the equivalent model (hr10) without the covariate included, this confirms that the results from the two groups of observers can be combined.

The hr10 model proves to give a lower AIC value than any alternative. It gives the 95% confidence interval for the number of nightingale reed warblers on Alamagan as being between 640 and 2090, with a best estimate as about 1160.

8.6   Multiple species

Since sampling requires time, effort, and expense, it makes sense to collect information on several species simultaneously. Some species are large and quite easy to detect (for example, herons), while others are small and quite difficult to detect (for example, quails). Such different species might have very different detection functions, and would therefore probably need to be analysed separately.

Suppose, however, that there are two species (one common, one scarce) that are of a similar size and are equally easy to detect. These may have very similar detection functions that differ only by a scaling factor that reflects the difference in their abundance. If the data for the scarce species were treated separately, then the rather few observations might lead to an uncertain estimate (a wide confidence interval). However, if the data for the scarce species is analysed in conjunction with that for other species (treating 'species' as a covariate), then a more precise estimate may be obtained (though this is not always the case).

Example 8.5: Yellow-fronted canaries in Hawai'i

In Hawai'i, the yellow-fronted canary (*Crithagra mozambica*) was scarce (97 birds in the 2016 survey) compared to the ubiquitous Japanese white-eye (1186 birds). The two species are of a similar size and colour, suggesting that it might be reasonable to expect similar detection functions. An initial analysis of the canary data on its own gives an estimated abundance of 0.57 per ha, with a 95% confidence interval being (0.38, 0.87).

To fit a detection function to the combined data from the two types of bird requires similar commands to those previously given for the white-eye data:

```
w<-which(data2016[,6]!="YFCA"&data2016[,6]!="JAWE")
data2016[w,5]<-NA; data2016[w,6]<-NA

library('distance')

jyhr75T<-ds(jyDF, transect="point", key="hr", adjustment=NULL,
truncation=75, formula=~Type)

plot(jyhr75T, pdf=TRUE)
```

The data have been summarized in the jyDF dataframe, with the vector Type holding the species information. After investigation, the combined distance data were truncated at 75 m. The resulting plot in Figure 8.9 shows that, on average, canaries were detected at greater ranges than white-eyes.

To obtain the separate estimates for the two species requires the following code:

```
ests <- dht2(ddf=jydshr75T, flatfile=jyDF, strat formula =
~Type, stratification = "object")
print(ests, report="density")
```

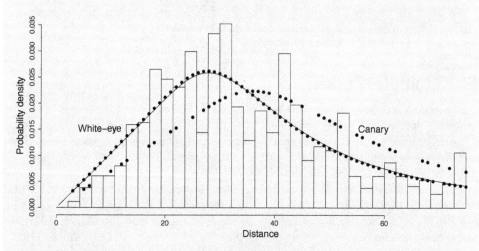

Figure 8.9 The density function using the jyhr75T model. White-eyes were detected at an average distance of 36.5 m, whereas canaries were detected at an average of 46.5 m.

The output includes the following:

```
Summary statistics:
  Type Area CoveredArea Effort    n    k
  JAWE    1    394.0735    223 1147 223
  YFCA    1    394.0735    223   85 223

Density estimates:
  Type Estimate    se    cv    LCI    UCI
  JAWE    8.6451 0.442 0.051 7.8202 9.5570
  YFCA    0.4203 0.084 0.200 0.2849 0.6201
```

The summary states that there were 223 sampling points where a total of 1147 white-eyes and 85 canaries were detected.[8] The estimated density for the canaries is 0.42 per ha, with a 95% confidence interval from 0.28 to 0.62 per ha. This is a considerably narrower interval than that obtained using the canary data alone.

8.7 Sightings of groups

In this chapter all the examples have referred to observations of singleton birds, but, when it comes to animals, sightings are more likely to involve family parties or larger groups.

A simple analysis that consists of treating each group as a singleton, obtaining an estimate of group abundance, and then scaling up by the observed group size, is likely to give an overestimate of the number of individuals present. This is because of the high proportion of undetected groups that consist of singletons or contain just a few individuals. Since, at any given distance, the probability of detection is affected by the size of the group, the solution is to treat group size as a covariate.

This applies equally to point transects and to line transects (there is an example in the next section).

8.8 Line transects

The analysis for line transects is very similar to that for point transects, but with one notable difference: the distance recorded is not the distance between the individual and the observer; it is the distance between the individual and the transect line (see Figure 8.10).

For a homogeneous population, the number of individuals that are present at (for example) a distance of between 5 m and 10 m from the transect line, will be the same, on average, as the number between 10 m and 15 m from the line. Indeed, it will be the same, on average, as in any strip of width 5 m that is parallel to the transect line. This means that the histogram of observed distances will not be humped, as was the case for a point transect, but will have the same form as the detection function.

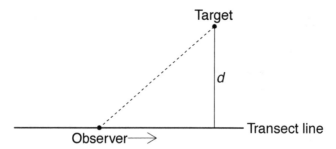

Figure 8.10 When using distance methods with a line transect, the distance required is the perpendicular distance from the transect line.

Example 8.6: Dall's sheep in Alaska

Schmidt and Rattenbury (2017) present data concerning aerial surveys of Dall's sheep taken in the Itkillik preserve subarea of the Gates of the Arctic National Park and Preserve in northern Alaska. In 2009 there were 26 randomly selected 20 km contour transects, with the plane travelling at a height of about 90 m. Figure 8.11 illustrates the results: in just over half the transects there were no sheep spotted, but on two transects there were seven groups identified. Most of the groups observed were relatively close to the transect line.

   One feature of the data that was scarcely apparent from Figure 8.11, but shows clearly in Figure 8.12, is the comparative absence of sightings close to the transect line. This is not uncommon with line transects, since the movement of the observer may cause individuals to move away from the line. With aerial surveys, there is the further consideration that the plane's body may restrict vision. Allowance is easily made by left-truncation of the data (so that any observations closer than some specified distance will be ignored). There may also be right truncation to eliminate observations at distances where the probability of detection falls below 15%.

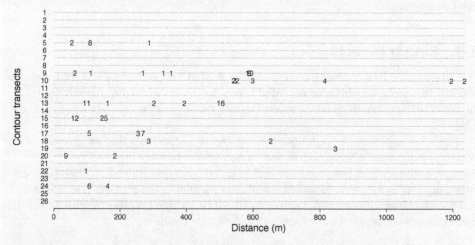

Figure 8.11 The numbers of sheep observed, and the distance of each group from the transect line, for 26 constant altitude contour transects flown over the Itkillik preserve in 2009.

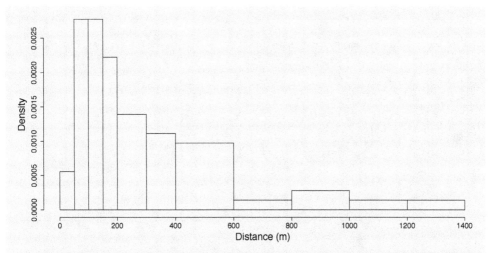

Figure 8.12 Histogram of the distances at which sheep groups were observed from the plane. Note the low frequency for the first distance category.

Figure 8.12 suggested that the data should be left-truncated at 50 m. After examining the graphs of fitted detection functions, the data were right-truncated at 700 m. The two detection functions with the lowest AIC values were hnct7 (a half-normal key with a cosine adjustment) and hrt7 (a hazard-rate key with no adjustment). Some relevant commands are given below:

```
library('distance')

sheep<-data.frame(Region.Label, Area, Sample.Label, Effort,
size, distance)

hrt7<-ds(sheep,transect="line", key="hr", adjustment=NULL,
truncation=list(left=50,right=700))

hnct7<-ds(sheep,transect="line", key="hn", adjustment="cos",
order=2, truncation=list(left=50,right=700))
```

The AIC value is a useful guide, but it is not a substitute for common sense. The fitted detection functions should be examined for signs of lack of fit that can be formally tested using the Cramér-von Mises test on the q-q plot. Results are illustrated in Figures 8.13 and 8.14.

Neither the elbow at 350 m for hnct7, nor the relatively low detection probabilities for hrt7 at distances of 50 to 150 m seem entirely plausible. However, the q-q plots show reasonable fits with tail probabilities in excess of 80%. It appears that either detection function could be used, though their abundance estimates differ considerably (see Table 8.3).

The uncertainty posed by the conflicting results for the models hnct7 and hrt7 suggests looking for relevant other information. In this case there is plenty available, since the surveys in this area continued annually until 2016. Studying the pooled data suggests truncation at 500 m rather than 700 m and the best-fitting detection function, ahnt5, uses the half-normal key with no extension. The following commands provide estimates for each year (with those for 2009 included in Table 8.3):

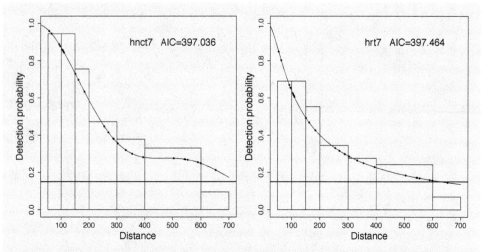

Figure 8.13 The two detection functions (without covariates) having the lowest AIC values. Neither are totally convincing.

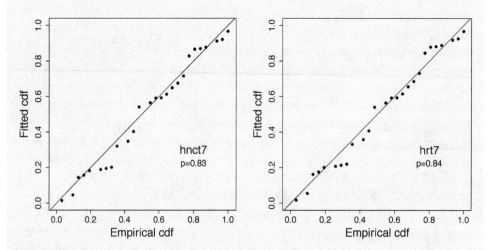

Figure 8.14 The q-q plots for the two detection functions: both are judged to be acceptable fits.

```
ahnt5<-ds(allsheep, transect="line", key="hn",
truncation=list(left=50,right=500))
yearests <- dht2(ddf=hnt5, flatfile=allsheep, strat formula = ~
year,
stratification = "geographical")
```

There remains the question of whether allowance should have been made for the variation in the numbers of sheep in the observed groups.

Table 8.3 Results from various models for the abundance of Dall's sheep in the Itkillik preserve in 2009.

Model	Key	Extension	Truncations (m) Left	Right	Lower 95% limit	Point estimate	Upper 95% limit
Using data from 2009 alone							
hnct7	Hn	Cosine	50	700	565	1308	3028
hrt7	Hr	None	50	700	181	1789	17,639
Using data from 2009 to 2016							
ahnt5	Hn	None	50	500	426	1003	2358

Key: Hn = Half-normal; Hr = Hazard rate

Figure 8.15 suggests that group size has not affected matters, since many small groups were seen at large distances. To test this requires the following R code:

```
ahnt5size<-ds(allsheep, transect="line", key="hn",
formula= size, truncation=list(left=50,right=500))
```

The result is an increase in the AIC value: thus, in this case, no allowance need be made for size. The final point estimate is therefore that, for this region, there were approximately 1000 Dall's sheep in 2009.

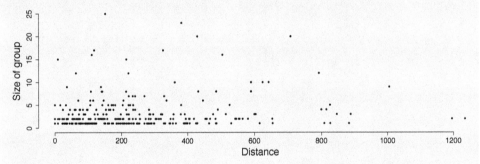

Figure 8.15 The plot of group size against distance shows little sign that the more distant groups are detected because of their larger size.

Part IV

Species

The previous chapters have dealt with methods for counting individuals, but in this final part the emphasis is on the number of species that are present. An ecosystem is applauded if there are many species present, but is a cause for worry if there are few species. A quick study of the literature reveals a bewildering profusion of rather similar and interrelated terms that are, to an extent, untangled in the following chapters.

Diversity, which is widely understood as referring to the extent of the variation in the species encountered in an ecosystem, is the subject of Chapter 10. However, 'diversity' is often regarded as an amalgam of 'richness' (the number of different species) which is the subject of Chapter 9 and 'evenness' (the extent to which the abundances vary from one species to another) which is discussed in Section 12.1. Other relevant words are 'equitability', 'dominance' and 'unevenness'. Whatever measures are used it is important to recognize their limitations: nature cannot be reduced to a few numbers!

9. Species richness

Species richness, S, is simply the number of different species present in the region being sampled.

The observed value of S depends on the sampling effort; in other words, the size of the area sampled, the time spent sampling, the sampling procedure, and so forth. These will rarely be uniform across populations, especially when examining possible changes over time. For this reason comparisons of the observed richness at different locations, or different times, require caution. Indeed, this warning applies to any of the measures discussed in this chapter. In any sampling effort, as the individuals are identified, so the number of species will increase. The increase is typically fast at first, but then slows, stopping when every possible species has been encountered. The graph of number of species, against sample size, is known as a *collector's curve*.

If the sample size increases as a result of increasing plot size, then the number of species may continue to increase, as a consequence of heterogeneity in the region sampled. The resulting plot of number of species encountered, against area sampled, is called a *species accumulation curve* (SAC).

Example 9.1: Indian trees

In the 2000 census of the trees in the Mudumalai Forest Dynamics Plot there were 12,574 individual specimens of at least 10 cm at breast height. These trees included specimens of 61 species.[1] Figure 9.1 is the average collector's curve resulting from 10 random samples taken without replacement. Notice how, after an initial period during which new species are regularly obtained, the curve levels off as new species become increasingly hard to find.

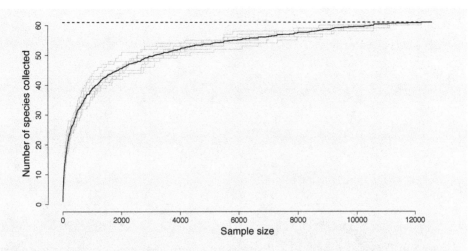

Figure 9.1 The average collector's curve resulting from 10 random orderings of the 12,574 individual trees in the 2000 census of the Mudumalai Forest Dynamics Plot. The dotted line indicates the total number of species present. Individual collections are shown in grey.

9.1 Richness indices

Since observed richness depends upon sample size, efforts have been made to standardize the results. Margalef (1958) suggested that 'by no means the worst' index of diversity was

$$M = \frac{S-1}{\ln(n)}, \tag{9.1}$$

where the denominator is the natural logarithm of the sample size, n. The quantity M is now referred to as the *Margalef richness index*.

Menhinick (1964) claimed that the Margalef index remained sensitive to the sample size, and proposed, as an alternative:

$$D = \frac{S}{\sqrt{n}}. \tag{9.2}$$

This is now referred to as the *Menhinick richness index*. While both M and D are simple to calculate, the next example demonstrates that (at least for small n) both are sensitive to the value of n.

Example 9.2: Indian trees (cont.)

Averaging over 10 random orderings of the Indian tree data, Figure 9.2 shows plots of M and D against n. It is apparent that the values of both indices are strongly affected by sample size at low sample sizes. For these data, it is M that most convincingly approaches some constant value. It is tempting to imagine examining the trees in an extended Indian plot with a million trees. In that case, using Equation (9.1), one might expect to have seen a further 27 species.

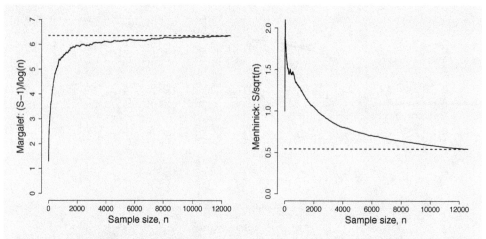

Figure 9.2 Plots of the Margalef and Menhinick richness indices, against sample size, for the Indian tree data. The curves shown are averages over 10 random samples. The limiting values are indicated by the dotted lines.

9.2 Rarefaction

This is an appealing method of dealing with comparisons of richness, at sites where different numbers of observations were obtained. Suppose that at one site, N individuals are observed including S_N species, with N_i individuals of species i. At the other site only n ($< N$) individuals are observed including S_n species. If $S_n < S_N$ then a question of interest is whether this is a consequence of the smaller number of observations.

If n individuals are randomly selected from the N individuals in the larger sample, then the average number of species obtained is

$$\sum_{i=1}^{S_N}\left\{1-\binom{N-N_i}{n}\Big/\binom{N}{n}\right\},\tag{9.3}$$

where

$$\binom{N}{n}=\frac{N!}{n!\times(N-n)!},\qquad n!=n\times(n-1)\times(n-2)\times\cdots\times1,$$

and $\sum_{i=1}^{S_N}$ denotes the sum over values of i from 1 to S_N.

To judge whether S_n is significantly less than S_N, an informative procedure is to take a large number of random samples of size n from the group of N individuals. For each sample, the statistic of interest is the number of species observed. If S_n is atypical of these numbers, then this will suggest an interesting result.

As with all comparisons of samples from different populations, a comparison can only be truly reliable if the same sampling procedure (i.e. the same sample method, the same time spent, etc.) is used. When this is not the case, any differences observed may be the result of different sampling techniques, rather than true differences in population composition.

Example 9.3: Indian trees and Californian trees

In 2000, there were 61 tree species at the Indian site, compared to the 31 species in the much smaller 200 m × 300 m UC Santa Cruz Forest Ecology Research Plot in California. If the latter had been extended to the size of the Indian site, would it have included as many species? The answer is unknown, but rarefaction can be used to examine the inverse question of how many Indian species might have been observed, if the sample had consisted of just 8372 trees (the number in the Californian site) rather than the 12,574 actually observed.

Figure 9.1 suggested that after examining over 8000 trees almost all the species will have been obtained. This is borne out by the results in Table 9.1 which summarizes the numbers of species observed in each of 999 random samples of 8372 trees taken from the 12,574 at the Indian site. Since 31 is far fewer than the smallest value observed, it can reasonably be concluded that the Indian forest is much richer than its Californian counterpart.

Figure 9.3 reports results from 100 random orderings of the 12,574 Indian trees. It shows that, on average, the investigator would need to see just 466 randomly chosen Indian trees[2] in order to have encountered 31 species.

Table 9.1 The numbers of species found in 999 samples of size 8372 taken from the 12,574 trees observed in the 2000 census of the Indian trees.

Number of species	52	53	54	55	56	57	58	59	60	61
Frequency of occurrence	1	3	14	43	124	231	266	199	99	19

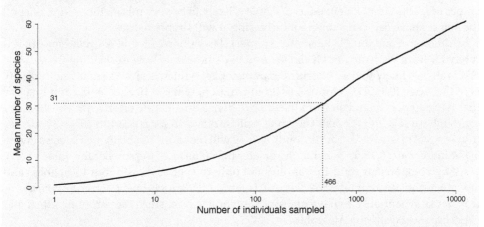

Figure 9.3 The average number of species seen as the number of observations increases. Averages based on 100 random rearrangements of the 12,574 trees recorded for the Indian plot.

9.3 The dependence of richness on area

The collector's curve (Figure 9.1) illustrates the fact that, as one increases the size of one's study, so S (the number of species encountered) will, in general, increase, and certainly cannot decrease. It is also evident that the rate of increase will generally slow down as the number of species already collected increases. The exception to the slowdown of accumulation occurs when the increase in area results in the study examining a different ecosystem. For example, if a survey extends from woodland to include a neighbouring lake, then woodland bird species will be supplemented by waterbirds and a second collector's curve will become attached to the first.

The first to propose a *SPecies-ARea (SPAR) model* was Arrhenius (1921) who observed that, for his data, S appeared to be related to A (the area sampled) by the simple power law:

$$S = \theta A^{\beta}, \tag{9.4}$$

where θ and β are constants. Equivalently, taking logarithms gives a convenient straight line relationship:

$$\ln(S) = \alpha + \beta \ln(A), \tag{9.5}$$

where $\alpha = \ln(\theta)$.

Palmer and White (1994) described the species-area curve as 'One of the most studied relationships in all of ecology' and gave a comprehensive review of its relevance in biogeography and related disciplines.

In the assessment of the relative biodiversity of very different regions (such as forests and heathlands), it is natural to compare their species richness. However, if one sampled region has a much greater area than the other, then some procedure is required to compensate for the difference in areas. Although a natural choice is to compare the values of *species density*, defined as S/A, this is likely to be very misleading, since, once all the species have been encountered, increasing A will simply reduce S/A.

Although Equation (9.4) is never a perfect description of the SPAR relationship, it is rarely far from the truth, and it can be used to demonstrate how misleading the value of S/A is always likely to be. The critical parameter is β, which is always small and typically is in the range (0.1, 0.3). Suppose, as an illustration, that $\theta = 10$ and $\beta = 0.3$ and imagine that two samples are taken from the *same* region: a small-scale effort with $A = 40$ and a thorough survey with $A = 500$. The small-scale survey will see about $10 \times 40^{0.3} \approx 30$ species, to give a S/A ratio of 0.75. The thorough survey will see about $10 \times 500^{0.3} \approx 65$ species giving a S/A ratio of just 0.13. In general, the greater the value of A, the smaller the value of S/A.

Any comparisons (and this applies not only to species density but to richness and any other statistics of interest) *should only be made if all the background factors are comparable.* It is not sufficient only to have areas of comparable size, since the sampling effort and expertise must also be comparable.

9.3.1 The relevance of the shape and orientation of the area sampled

In general, species are not randomly scattered across the area of interest, but occur with varying intensity from one location to another. This, in part, reflects the simple fact that neighbouring regions are similar to one another, whereas distant locations can be very dissimilar. The consequence is that a narrow rectangle of dimensions $n^2 \times 1$ is likely to encounter a greater number of species than a square area with dimensions $n \times n$.

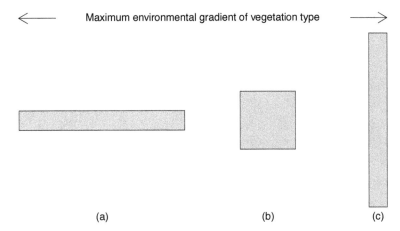

Figure 9.4 Alternative sampling plots, each having the same sampling area. The number of species observed using (a) is likely to exceed that obtained using (b), while the number obtained using (c) may be the least of the three.

Orientation is also relevant: for example, a rectangular region in the direction of steepest slope on a hillside, is likely to encounter more species than one that follows the contours (see Figure 9.4).

Example 9.4: Californian trees

Figure 9.5 illustrates the effect of plot shape and separation on the number of species encountered for the Californian data. The curves shown in Figure 9.5 result from fitting the power law given by Equation (9.4) to the average numbers of species observed in two alternative sequences of non-overlapping plots: square plots with sides ranging from 5 m to 40 m, and rectangular plots with a short side of 5 m and a long side ranging from 10 m to 200 m. The greater number of species using rectangular plots is immediately apparent.

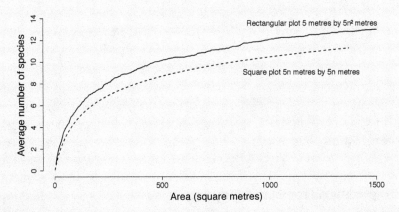

Figure 9.5 The number of species encountered in a narrow rectangle generally exceeds the number encountered in a square region of the same area. This is illustrated for the Californian trees.

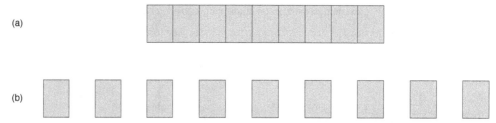

Figure 9.6 Alternative arrangements of nine sampling plots. The number of species found using contiguous plots, (a), is likely to be less than the number found using non-contiguous plots, (b).

9.3.2 The relevance of separation in sampled locations

Positive spatial autocorrelation implies that adjacent locations have similar characteristics. Heterogeneity in species is therefore achieved by sampling locations that are distant from one another (see Figure 9.6).

> ### Example 9.5: Californian trees (cont.)
>
> Figure 9.7 demonstrates the contribution to the increased number of species, that results from increasing the distance between the locations being sampled. Similar results are presented in a more extensive study by Palmer and White (1994).
>
>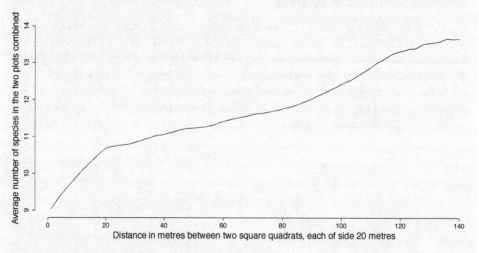

Figure 9.7 On average, the total number of species seen using two plots increases with the distance between the plots.

9.3.3 Implications for island biogeography

In the current context, an island refers either to a piece of land surrounded by sea, or to a habitat island. The latter might be a remnant of natural landscape surrounded by areas devoted to habitation or agriculture. Since locations that are distant from one another, are more likely to result in the discovery of new species, than are neighbouring locations, the *island species-area relationship (ISAR) curve*, formed by accumulating information from separate islands, will be steeper than the curve resulting from equivalent species-area data obtained from subregions of a connected region. An in-depth discussion of the differences between SAC and ISAR curves is provided by Matthews et al. (2016).

Example 9.6: Californian trees (cont.)

Figure 9.8 shows the species present in one 40 m × 40 m corner of the plot. The results are typical of those seen, not just for Californian trees, but for organisms in general. The features are:

- The density of individuals varies considerably across the region studied. This is due in part to variations in the local environment, but also reflects the sizes of the individuals concerned.

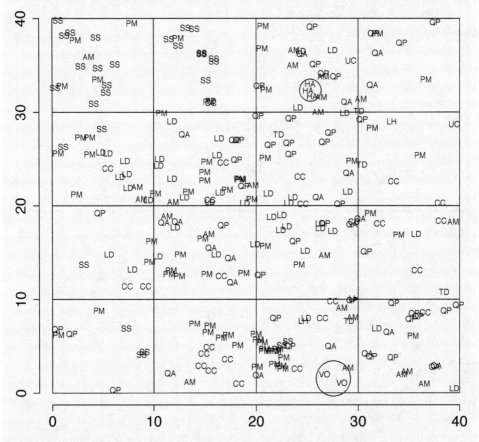

Figure 9.8 The trees present in a 40 m × 40 m corner of the Californian tree plot. The species *Pseudotsuga menziesii* (PM) appears in each of the 10 m square sections illustrated, but two species *Heteromeles arbutifolia* (HA) and *Vaccinium ovatum* (VO) appear only in a single 10 m square, with each present as a tight group.

- Some species are abundant and seemingly ubiquitous. An example here is *Pseudotsuga menziesii* (PM) which appears in each of the 10 m square sections illustrated.
- Other species are scarce, with some appearing only in small subregions. Examples here are *Heteromeles arbutifolia* (HA) and *Vaccinium ovatum* (VO) with small clumps indicated by circles on the diagram.

Figure 9.9 indicates the positions of 12 non-overlapping regions of the Californian tree plot. The density of trees evidently varies across the plot and this is reflected in the locations of individual tree species.

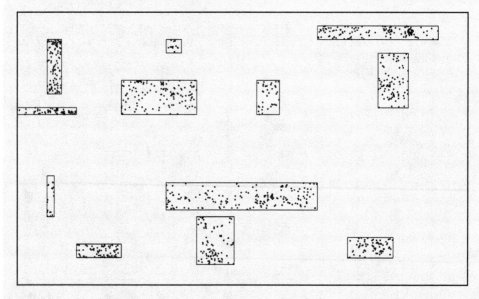

Figure 9.9 Twelve artificial islands used within the 200 m × 300 m Californian forest plot.

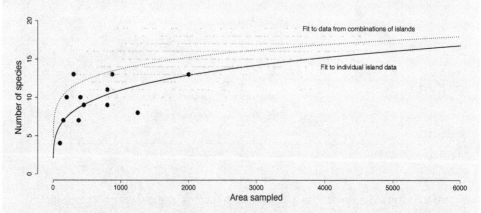

Figure 9.10 The large black points refer to the 12 artificial islands illustrated in Figure 9.9. The small grey points refer to combinations of islands. The continuous line is the power-law fit to the island data. The dotted line is the power-law fit to the combination data.

Figure 9.10 shows the power-law curve fitted to the individual island data. No island contained more than 13 species. It suggests that an unseen island with an area of 6000 m^2 might contain 14 species. Also shown is the power-law curve fitted to aggregates of the island data. This suggests that combining islands to arrive at an area of 6000 m^2 might result in observation of 19 species.

The results of Matthews et al. (2016), who studied the incidence of spider species in genuine habitat islands, show that the difference between the two power-law fits can be much greater than that shown here.

---

❶ *How the data were obtained*

To determine the positions of the trees, the surveyors worked with 20 m squares. For each square the first task was the establishment of the four corner posts in their correct positions. Subsequently, for each tree within a square, a laser was used to measure the distances to all visible corner posts. Compass bearings were also taken. The information was combined using elementary trigonometry to establish the tree position within the 20 m square and hence its overall position in the survey area.

9.4 Estimating the unobserved

In an influential paper, Colwell and Coddington (1994), the authors remarked: 'The urgent challenges of global climate change, massive habitat transformation, and the threat of widespread extinction, … have made extrapolation and prediction a crucial component of many research agendas [in the study of biodiversity].' The subsequent literature contains a considerable discussion on this subject, with general agreement that the most plausible estimators are two pairs of estimators introduced by Professor Anne Chao and her co-workers.

During World War II, in the context of cryptographic analysis, Good and Turing (reported by Good (1953)) had found that a good estimate of the proportion of items that were of types as yet unseen was given by f_1/n, where n is the total number of items seen, and f_1 is the number seen only once. The Good-Turing approach was further developed in Chao (1984, 1987) and Chao and Lee (1992), with freely available computer programmes that provide both values, and confidence intervals, for their estimates of the number of unseen species. However, the examples that follow, demonstrate that increases in the sample size persistently lead to numbers of species that exceed predictions.

9.4.1 Chao1 and Chao2

These estimators were introduced in Chao (1984, 1987). Let n be the total number of individuals, let f_k denote the number of species that are represented by exactly k individuals, and denote the estimator introduced by Chao (1984) as $\widehat{S_C}$. With the modification for the case $f_2 = 0$ that is used by Chao and Shen (2010) in their *SPADE* programme, $\widehat{S_C}$ is given by[3]

$$\widehat{S_C} = S_{obs} + \begin{cases} \frac{n-1}{2n} \frac{f_1^2}{f_2}, & f_2 > 0; \\ \\ \frac{n-1}{2n} f_1(f_1 - 1), & f_2 = 0. \end{cases} \tag{9.6}$$

Chao (1987) showed that the variance of $\widehat{S_C}$ was

$$\frac{1}{4} f_2 (r^4 + 4r^3 + 2r^2), \tag{9.7}$$

where $r = f_1/f_2$.

A slight adjustment to the original formula is provided by incorporating an extra term (Chiu et al., 2014). The result, $\widehat{S_{CC}}$, is given by

$$\widehat{S_{CC}} = \widehat{S_C} + \begin{cases} \frac{n-3}{4n} \frac{f_3}{f_4} \times \max\left(0, f_1 - \frac{n-3}{2(n-1)} \frac{f_2 f_3}{f_4}\right), & f_4 > 0; \\ \\ \frac{n-3}{4n} f_3 \times \max\left(0, f_1 - \frac{n-3}{2(n-1)} f_2 f_3\right), & f_4 = 0. \end{cases} \tag{9.8}$$

For many organisms counting the number of individuals present may be infeasible or impractical, though the presence or absence of a species can be recorded. For these *incidence*-based cases, the previous estimators can be used with small changes to the definitions of n and the f_k-values. For the incidence case, n is then taken to be the number of samples, and f_k is the number of species that occur in exactly k samples. When used with frequencies, $\widehat{S_C}$ is referred to as the *Chao1* estimator; when used with species incidence it is called the *Chao2* estimator. The calculations are easily achieved using either the *iNEXT* or *SPADE* programmes.[4] For example:

```
# The species abundances are in the vector counts
library(iNEXT)              # Calls the appropriate R library
iNEXT(counts)               # Reports the observed and estimated
                            # richness, with 95% limits
```

In a subsequent paper, Chao et al. (2015) developed a means of estimating the entire rank-abundance distribution from that part actually observed. The procedure is complex, but an R programme, *JADE*, is freely available.

Example 9.7: Costa Rican ants

Longino, Coddington, and Colwell (2002) examined the richness of the ant fauna collected at the La Selva Biological Station in Costa Rica. Ants were trapped by several methods. The results given here result from 18 trees that were canopy fogged in 1993–1994, and 6 trees that were fogged in 1999–2000.[5]

Figure 9.11 shows the results of the *SPADE* programme based on the cumulating information from 459 fogging samples. The first sample contained 20 species giving an estimate of 35 unseen species. That predicted total of 55 species had been encountered by sample 17, at which point the estimator predicts the existence of a further 16 species. Similarly, after 71 species the prediction is for a further 27; after 98 species the prediction is for a further 35; after 133 species a further 26; after 159 species a further 14. After the 459th sample, a total of 164 species had been observed, with a prediction of a further 19 and an associated confidence interval of between 8 and 50 further species.

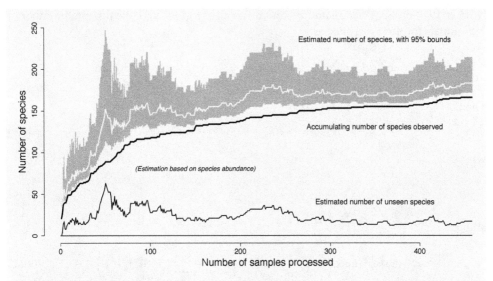

Figure 9.11 Estimated numbers of ant species obtained by fogging in Costa Rica. The estimate (white line) is shown with the 95% confidence interval (in grey), the number of species (thick black line) and the estimated number as yet unseen (thin black line). There were 459 samples collected in 1993/4 and 1999/2000. The estimated confidence intervals are those given by the *SPADE* programme.

Note that, however believable the prediction, it is limited to the trapping method used at the location specified. The total number of species in the vicinity is possibly three times the number obtained using fogging since Longino, Coddington, and Colwell (2002) report having found ants from a total of 455 species at La Selva.

---

❶ *How the data were obtained*

For canopy fogging, forty 1 m² funnels were suspended beneath the crown of a canopy tree, the crown was fogged with synthetic pyrethrins, and the funnels were harvested after a 2-hour drop time. Most ants were identified in alcohol, with point-mounting of those requiring it for identification. A detailed account of canopy fogging is given by Erwin (1983).

9.4.2 ACE and ICE

These estimators, which were introduced by Chao and Lee (1992), divide the species into two groups, one labelled rare (represented by k individuals) and the other labelled abundant (represented by more than $\leq k$ individuals). The user guide for the *SPADE* programme suggests using $k = 10$ unless the estimated number of species is less than that obtained using the Chao1 estimator, $\widehat{S_C}$, in which case a larger value of k is suggested. A larger value is also suggested if species have 'very unequal detection probabilities'.

The estimators are S_{ACE}, the Abundance-based Coverage Estimator, and S_{ICE}, the Incidence-based Coverage Estimator. For S_{ACE} and the chosen value for k, let $S_R(k)$ and

$S_A(k)$ be the numbers of rare and abundant species, and let $n_R(k)$ be the total number of individuals for the $S_R(k)$ species:

$$S_R(k) = \sum_{i=1}^{k} f_i, \qquad S_A(k) = \sum_{i>k} f_i, \qquad n_R(k) = \sum_{i=1}^{k} i f_i.$$

Then

$$S_{ACE} = S_A(k) + \theta\{S_R(k) + f_1 C(k)\}, \tag{9.9}$$

where

$$\theta = \frac{n_R(k)}{n_R(k) - f_1}, \tag{9.10}$$

and

$$C(k) = \max\left(0, \frac{S_R(k)}{[n_R(k) - f_1][n_R(k) - 1]} \sum_{i=1}^{k} i(i-1)f_i - 1\right). \tag{9.11}$$

Denoting the number of samples that include at least one rare species by $R(k)$, the same equations apply for S_{ICE}, with definitions now referring to the numbers of samples within which species were detected, and with an extra term, $R(k)/\{R(k) - 1\}$, included in the product in Equation (9.11).

Example 9.8: Costa Rican ants (cont.)

Table 9.2 shows the values of f_i, for $i = 1, 2, \ldots, 10$. With $k = 10$, $S_R(k) = 19 + 13 + \cdots + 4 = 68$, and $n_R(k) = 19 + 26 + \cdots + 40 = 250$. Since a total of 164 species were observed by this method, $S_A(k) = 164 - 68 = 96$. Also $\theta = 250/231 = 1.08$ and

$$C(k) = \frac{68 \times (0 + 26 + 48 + \cdots 360)}{231 \times 249} - 1 = 0.383,$$

so that $S_{ACE} = 96 + 1.08 \times (68 + 19 * 0.383) = 177.5$. The estimate provided by this procedure is very close to the 177.9 given previously by the Chao1 estimator.

The freely available *SpadeR* programme calculates estimates and confidence intervals for S_C, S_{ACE} and several alternatives and provides associated confidence intervals. The commands are simple:

```
# The species abundances are in the vector counts
library(SpadeR)              # Calls the appropriate R library
ChaoSpecies(counts)
```

The accumulating graph of observed number of species and associated predictions closely resembles that seen previously in Figure 9.11; the two estimators appear equally plausible. In this case, the choice of value for k makes little difference with $k = 1$ giving an estimate of 177.9 and $k = 20$ giving 181.1.

Table 9.2 Numbers of species represented by $i = 1, 2, ..., 10$ individuals.

Number of individuals, i	1	2	3	4	5	6	7	8	9	10
Number of species, f_i	19	13	8	4	7	6	4	1	2	4

9.5 The limitation of using richness as a measure of diversity

Knowing the number of species present is certainly interesting, but it is far from a complete description of the variety of species present. This is illustrated by Table 9.3, which presents the results for two equal-sized sample regions that contain the same numbers of individuals and species. The first sample contains one dominant species and four rare species, whereas the second sample contains five equally abundant species. Few observers would consider them to be equally diverse.

Table 9.3 Two samples with the same size ($n = 20$ organisms) and the same species richness ($S = 5$). Few observers would consider them to be equally diverse.

A	A	A	A	A	A	B	C	D	E
A	A	A	A	A	A	B	C	D	E
A	A	A	A	A	A	B	C	D	E
A	B	C	D	E	A	B	C	D	E

9.6 An occupation-detection model

Guillera-Arrolta, Kéry, and Lahoz-Monfort (2019), suggest that the number of species present, may be estimated by modelling the data from repeat visits to sampling sites within the region of interest. Their approach echoes that used in modelling capture-recapture data.

At each sampling site, a species is either present or absent. If it is present, then it is either detected, or it is not detected. The data consist of a set of y-values, each equal to 0 or 1, with the value of y_{ijk} recording whether or not species k was identified at site i on the jth occasion. The model assumes that the question of whether or not species k is present at site i, can be regarded as an observation, z_{ik}, from a Bernoulli distribution (see Section 1.4.1) with *occupancy probability* ψ_{ik}. If the species is present at that site, then $z_{ik} = 1$, and if it is absent then $z_{ik} = 0$. The value of ψ_{ik} is likely to vary from site to site, and this variation may be captured using a logit model:

$$\log\left(\frac{\psi_{ik}}{1 - \psi_{ik}}\right) = \beta_{0k} + \beta_{1k}X_{1i} + \beta_{2k}X_{2i} + \cdots,$$

where $X_{1i}, X_{2i}, ...,$ are characteristics of site i.

The next question is whether or not the species was seen on visit j. For a species k that is present at site i, let p_{ijk} be the probability that it is detected on the jth visit. Then y_{ijk} can be considered as an observation from a Bernoulli distribution with parameter $z_{ik}p_{ijk}$. The value of p_{ijk} could be linked, using a logit model, to features of the sampling occasion, such as weather conditions.

With $\widehat{\psi}_{ik}$ denoting the estimate of ψ_{ik} obtained using this model, an estimate of the number of species present at site i is

$$\sum_k \widehat{\psi}_{ik}.$$

To estimate the number of unseen species requires further elaboration of the model by assuming that the β-parameters are linked across species. The link takes the form of assuming that each β is an observation from a normal distribution. So, for example, β_{0k} is assumed to be an observation from a normal distribution with mean μ_0 and variance σ_0^2.

The next stage in the estimation process is to select a value for M, the maximum number of species that could conceivably be present in the study region. Each of these hypothetical M species may or may not be present: so there is yet another Bernoulli random variable, w_k. For any species that is present (but may be undetected at every site) $w_k = 1$; otherwise it is equal to zero. The authors suggest that the model performs well in situations where most species present have been detected.

10. Diversity

Somewhat confusingly, Whittaker (1972) wrote that 'Diversity in the strict sense is richness in species, and is appropriately measured as the number of species in a sample of standard size.' He went on to describe this as α-diversity. Locations in the neighbourhood will not be identical, so Whittaker described the 'extent of differentiation of communities along habitat gradients' as β-diversity. Whittaker described the product of α-diversity and β-diversity as γ-diversity, which might be described as the overall richness and complexity of the local environment.

The most used measures of α-diversity are probably Berger-Parker dominance, Shannon entropy and the Simpson index, which are the subjects of the next sections.

10.1   Berger-Parker dominance

Suppose that a sample consists of n individuals, with S species being represented. Let n_i be the number of individuals of species i. Let the species be identified so that $n_1 \geq n_2 \geq \cdots \geq n_S$. The very simple measure suggested by Berger and Parker (1970) is:

$$D = \frac{n}{n_1},\qquad\qquad(10.1)$$

Variants using $(n_1 + n_2)$ or $(n_1 + n_2 + n_3)$ have also been suggested.

10.1.1   The *k*-dominance curve

Berger-Parker dominance uses information only from the single most dominant species. By contrast, the k-dominance curve provides information for the k most abundant species. It is a plot of the rank of a species against the cumulative proportion of individuals having that, or a lower rank. The extent of the dominance of the leading species will be reflected in the extent to which the curve deviates from the straight line that would result if every species was equally frequent.

To give more visibility to the most abundant species, the value of k, on the x-axis, is often presented using a logarithmic scale.

Example 10.1: Indian trees (cont.)

Figure 10.1 displays the k-dominance plot for all the species represented in the Indian tree data for 2000. The use of a log scale for the x-axis barely helps with visibility of the data.

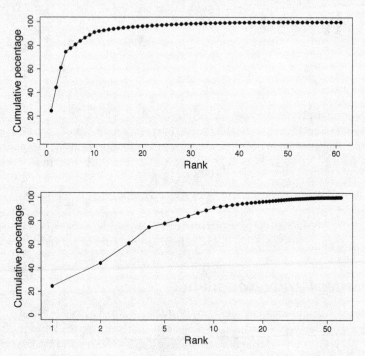

Figure 10.1 *K*-dominance plots for the Indian tree data (2000 count). The lower diagram uses a log scale for ranks.

10.1.2 Partial dominance

When, as is usually the case, there is one (or more) very dominant species, then the k-dominance curve is distorted in a fashion that makes it difficult to observe the contributions from the remaining species. A possible solution suggested by Clarke (1990) is to plot *partial dominance*, defined as the proportion of the remaining individuals that are accounted for after the more dominant species have been excluded.

Example 10.2: Indian trees (cont.)

The single most dominant species accounted for nearly a quarter of the 12,574 individual trees counted in 2000. Eliminating this species from consideration leaves 9482 trees, of which 2482 (26%) are accounted for by the next most abundant species. The partial-dominance plot is given in Figure 10.2. The right-hand end of the plot results from single-figure counts and conveys little information. Whether the left-hand side is useful is a decision for the reader.

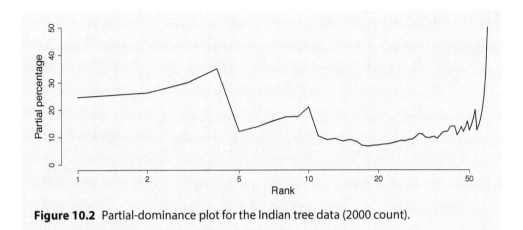

Figure 10.2 Partial-dominance plot for the Indian tree data (2000 count).

10.2 Shannon entropy

Shannon (1948) introduced the quantity, H, as a measure of uncertainty. In the current context this is uncertainty concerning the species type, when, at a given location, an individual is selected at random. With S different species present, of which a proportion p_i belong to species i, H is given by

$$H = -\sum_{i=1}^{S} p_i \ln(p_i). \tag{10.2}$$

If there is just one species present, then there is no uncertainty and $H = 0$. The maximum value of H is $\ln(S)$ which occurs when there are equal numbers for every species so that $p_i = 1/S$, for all i.

If a sample of n individuals contains n_i of species i, then replacing p_i by n_i/n in Equation (10.2), gives the sample estimate, $\hat{H}$. Chao, Wang, and Jost (2013) provide an adjustment that allowed for the fact that, since a sample is unlikely to include every species in the population, $\hat{H}$ will generally under-estimate H. The revised formula is very complex, but values may be obtained using the ChaoEntropy command of the *iNEXT* package.

For a completely enumerated population, Pielou (1979) showed that a more appropriate measure is

$$H_B = \frac{1}{n}\left(\ln(n!) - \sum_{i=1}^{S}\ln(n_i!)\right), \tag{10.3}$$

which is due to Brillouin (1962).

The value taken by Shannon entropy can be difficult to interpret. For example, suppose that, initially, there are 20 individuals comprising four each of five types. The species richness is 5 and the Shannon entropy is $\ln(5) = 1.61$. Now suppose instead that there are 40 individuals, with four each of ten types. The species richness doubles to 10, while the revised value for Shannon entropy is $\ln(10) = 2.30$. The increase from 1.61 to 2.30 is not easily interpreted as a doubling of the number of species.[1]

10.3 Simpson's index

Simpson (1949) proposed, as a very simple index of population concentration, the quantity λ. This is defined as the probability that two randomly chosen individuals belong to the same species:

$$\lambda = \sum_{i=1}^{S} p_i^2. \tag{10.4}$$

This has value 1 when there is a single species present, and the value $1/S$ when all the species present are equally common (so that $p_i = 1/S$ for all i). The sample estimate of λ, $\hat{\lambda}$, is given by:

$$\hat{\lambda} = \frac{1}{n(n-1)} \sum_{i=1}^{S} n_i(n_i - 1). \tag{10.5}$$

Since high diversity corresponds to small values of λ, the formula given may be referred to as *Simpson's index of dominance*, or simply *dominance*. The quantity $1 - \lambda$ is variously referred to as *Simpson's index of diversity*, the *Gini-Simpson index*, and the *probability of interspecific encounter*. The phrase *Simpson's index of diversity* may also refer to $1/\lambda$, which the next section suggests is a preferable index.

10.4 Effective numbers

Hill (1973) defined the quantity N_k by

$$N_k = \left(\sum_{i=1}^{S} p_i^k \right)^{1/(1-k)}, \qquad (k \neq 1). \tag{10.6}$$

If every species is equally likely, then $p_i = 1/S$ for all i. Substituting this value shows that, in that case, $N_k = S$ for all values of k.

The case $k = 0$ gives most weight to rare species, with N_k then having its maximum value, S.

Setting $k = 1$, and taking limits, gives

$$N_1 = \exp(H), \tag{10.7}$$

where H is the Shannon entropy given by Equation (10.2) and exp() is the exponential function.

Since, with S_c equally common species, $H = \ln(S_c)$, N_1 can be interpreted as the number of equally common species that would have resulted in the observed value of H. MacArthur (1965) described this as the effective number of species. For this reason, the $\{N_k\}$ are variously described as *effective numbers* or *Hill numbers*.

Setting $k = 2$, gives

$$N_2 = 1 \left/ \sum_{i=1}^{S} p_i^2 \right. = 1/\lambda \tag{10.8}$$

implying that $\lambda = 1/N_2$. Since, with S_c equally common species, $\lambda = 1/S_c$, N_2 can also be interpreted as an effective number of species.

Chao et al. (2014) present a procedure for estimating the values of the effective numbers that would have been obtained with a different sample size. The procedure is implemented in the freely available *SpadeR* programme.

As k increases so less attention is paid to the rare species and the values of N_k rapidly reduce to $1/M$, where $M = \max_i(p_i)$. Thus N_∞ is the reciprocal of D, the Berger-Parker dominance.

Example 10.3: Indian trees (cont.)

Figure 10.3 shows a plot of the effective numbers for the Indian forest data for 1988 and for 2000. The reduction in richness from 63 to 61 species is reflected in slight reductions of the effective numbers at all orders. As the plot shows, there is little change in N_k for $k > 2$, with the limiting values being 4.8 and 4.1. Since $63 > 61$ and $4.8 > 4.1$, the reduction in diversity is clear whichever value of k is chosen. If the relative magnitudes of the limiting values had been reversed then whether or not the forest had increased or decreased in diversity would have been unclear.

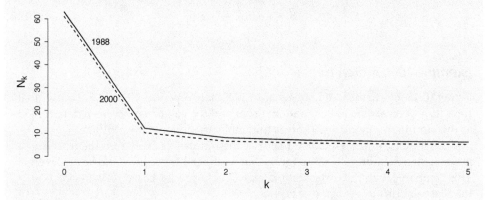

Figure 10.3 Plot of N_k against k for the Indian tree data. A slight reduction in diversity over time is apparent whichever measure is used.

10.5 Fisher's α

In Fisher, Corbet, and Williams (1943), Fisher suggested that, in large samples, the frequencies with which the various species occur can often be reasonably approximated by a log-series distribution (Section 1.5.1). Suppose that a sample of n individuals includes representatives of S species, Fisher showed that $E(S_k)$, the expected number of species having exactly k individuals, was given by:

$$E(S_k) = \frac{\alpha x^k}{k},$$
(10.9)

where x and α are the joint solutions of

$$S = -\alpha \ln(1 - x),$$
(10.10)

$$n = \frac{\alpha x}{1 - x}.$$
(10.11)

The larger the value of α, the greater the imbalance in the numbers of the species present. Since α is essentially independent of the sample size, Fisher suggested that α might be regarded as an index of diversity.

Rearranging Equation (10.11) gives

$$x = \frac{n}{n+\alpha}$$

and substitution for x in Equation (10.10) then gives

$$S = -\alpha \ln\left(\frac{\alpha}{n+\alpha}\right). \tag{10.12}$$

One simple approach to finding the solution of this equation is to evaluate the right-hand side for a sequence of possible values of α, choosing $\hat{\alpha}$ as the value that results in the expression being closest in value to S.

Having estimated the value of α, Equation (10.12) can then be used to estimate the number of species, $\hat{S}$, that might be discovered in a population of size N:

$$\hat{S} = -\hat{\alpha} \ln\left(\frac{\hat{\alpha}}{N+\hat{\alpha}}\right). \tag{10.13}$$

Example 10.4: Indian trees (cont.)

Figure 10.4 refers to the Indian tree data of 2000 (see Appendix). The 12,574 individual trees have been arranged in a random order and the value of α has been determined as the sample size increases. There is minimal variation in α for $n > 1000$.

If the plot could be increased so that the number of trees were increased to a million, then Equation (10.13) suggests that a further 37 species would be discovered. This compares with the earlier estimate of 27 new species based on extrapolation of the Margalef richness index (see Section 9.1).

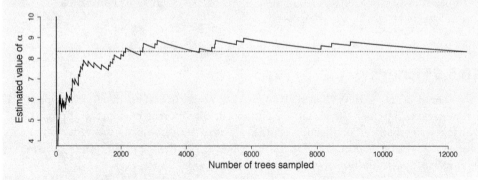

Figure 10.4 Plot of α against n for the Indian tree data. It appears that the value is nearly constant ($= 8.33$) when the sample size exceeds 1000 trees.

10.6 Taking account of differences between species

10.6.1 Quadratic entropy

Rao (1982) considered a number of measures including Q, defined by

$$Q = \sum_{i=2}^{S} \sum_{j=1}^{i-1} d_{ij} p_i p_j, \tag{10.14}$$

where $d_{ij} = d_{ji}$ for all i and j. The quantity $\{d_{ij}\}$ is a measure of the 'distance' between species i and species j. Rao (2010) showed that Q may be used to subdivide overall diversity into contributions within and between subpopulations. He noted that d_{ij} might be either quantitative or qualitative. If $d_{ij} = 1$ for all i and j, then $Q = \lambda$ (the Simpson index).

With sample data, Q would be estimated by

$$\widehat{Q} = \frac{1}{n(n-1)} \sum_{i=2}^{S} \sum_{j=1}^{i-1} d_{ij} n_i n_j, \tag{10.15}$$

where n_i is the number of individuals of species i, and n is the total number of individuals.

10.6.2 Taxonomic diversity

Warwick and Clarke (1995) suggested associating numerical values with the distances between pairs of individuals by reference to the extent of their taxonomic difference. For example, for individuals of distinct species belonging to the same genus, $d_{ij} = 1$; for individuals of different genera but the same family, $d_{ij} = 2$; for individuals of different families but the same order, $d_{ij} = 3$, and so forth. They defined two quantities closely related to $\widehat{Q}$, namely:

$$\Delta = \frac{2}{n(n-1)} \sum_{i=2}^{S} \sum_{j=1}^{i-1} d_{ij} n_i n_j, \tag{10.16}$$

and

$$\Delta^* = \sum_{i=2}^{S} \sum_{j=1}^{i-1} d_{ij} n_i n_j \bigg/ \sum_{i=2}^{S} \sum_{j=1}^{i-1} n_i n_j, \tag{10.17}$$

where n_i is the number of individuals of species i, n is the total number of individuals, and there are S species represented.

The quantity Δ is the average taxonomic difference (in terms of the scale used) between two randomly chosen *individuals* in the sample, while the quantity Δ^* is the average taxonomic difference between two randomly chosen *species* in the sample.

If the sample information simply reports the presence or absence of individual species, then both quantities reduce to Δ^+, given by

$$\Delta^+ = \frac{2}{S(S-1)} \sum_{i=2}^{S} \sum_{j=1}^{i-1} d_{ij}. \tag{10.18}$$

Clarke and Warwick (1999) showed that, when comparing samples from different populations, similar ratios for the Δ parameters were obtained for any sensible choice of values for the $\{d_{ij}\}$.

Example 10.5: Indian trees and Californian trees (cont.)

We use the scheme suggested above, namely for individuals of distinct species belonging to the same genus, $d_{ij} = 1$; for individuals of different genera but the same family, $d_{ij} = 2$; for individuals of different families but the same order, $d_{ij} = 3$.

For the Indian trees the values of Δ, Δ^* and Δ^+ are, respectively, 2.33, 2.75, and 2.939, while the corresponding figures for the Californian trees are 2.42, 2.82 and 2.942.

Thus, whichever measure is used, the Californian trees appear to be slightly more taxonomically diverse than the Indian trees.

10.6.3   Generalized effective numbers

Leinster and Cobbold (2012) wrote: 'Non-specialists are amazed to learn that a community of six dramatically different species is said to be no more diverse than a community of six species of barnacle. There is a mismatch between the general understanding of biodiversity as the variety of life, and the diversity indices used by biologists every day.' They suggested a measure that combines the ideas of effective numbers, and distances between species, into new quantities $\{M_k\}$, $(k \geq 0)$ given (for $k \neq 1$)

$$M_k = \left(\sum_{i=1}^{S} p_i E_i^{k-1} \right)^{1/(1-k)}, \tag{10.19}$$

where

$$E_i = \sum_{j=1}^{S} s_{ij} p_j, \tag{10.20}$$

and, in the case of counts, $p_i = n_i/n$, where n_i is the number observed for species i and n is the total number observed. The quantity s_{ij} measured on a scale from 0 to 1, is the similarity between species i and species j, with $s_{ii} = 1$. Thus E_i is the expected similarity between an individual of species i and a randomly chosen individual. Two special limiting cases are

$$M_1 \quad = \quad 1/(E_1^{p_1} \times E_2^{p_2} \times \cdots \times E_S^{p_S}) \tag{10.21}$$

$$M_\infty \quad = \quad 1/\max_{i=1}^{S}(E_i) \tag{10.22}$$

If $s_{ij} = 0$ for all $i \neq j$, then $M_k = N_k$, for all k. Thus M_k is the effective number of totally distinct species.

Leinster and Cobbold (2012) observe that diversity cannot be effectively encapsulated using a single number; they advise plotting the graph of M_k against k (they refer to this as a *diversity profile*).

An outstanding question is how to measure s_{ij}. This might be based on the extent to which species are similar in functional traits (*functional diversity*), DNA composition (*genetic diversity*), location on a phylogenetic tree, or some combination of these. Botta-Dukát (2005) suggests that a comparison based on multiple traits should use Mahalanobis generalized distance.

Example 10.6: Indian trees (cont.)

In this case, to calculate the M_k-values, s_{ij} is given the value 0.9 if i and j refer to different species from the same genus, with the value 0.7 for species from the same family, 0.5 for species from the same order, and 0.2 otherwise. These are an arbitrary choice of scores. The resulting profiles for 1988, 1992 and 2000 show an ever-decreasing diversity (see Figure 10.5).

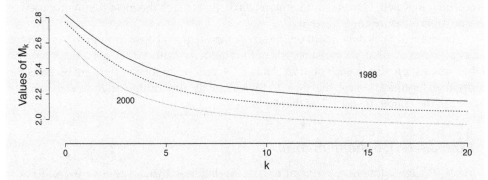

Figure 10.5 Generalized effective numbers for the Indian trees data. The three profiles, for 1988, 1992, and 2000, show a steady reduction in diversity.

10.7 Measuring β-diversity

The previous sections have concentrated on the measurement of α-diversity, which is an assessment of the intrinsic variability of species within a homogeneous environment. By contrast, β-diversity is intended to reflect the variation in species due to a change in environment. In the context of global warming the last phrase might read 'due to a changing environment'.

Koleff, Gaston, and Lennon (2003), compared 24 alternative measures of β-diversity. They observed that 'for each new application of the concept of beta diversity [it was almost the case that] a fresh measure has been derived'. It is apparent that comparisons of β-diversity between regions, or across time, will need to be based on the same measure of β-diversity and must be defined using comparable regions and subregions. They found that none of the 24 measures satisfied all the criteria that they considered desirable.

Anderson et al. (2011) examined 27 measures and concluded: 'Rather than choosing a single measure of β-diversity, we recognize that communities have a variety of ecological properties of interest, and we advocate using a suite of measures, each driven by specific hypotheses.' In this section, only the most widespread and apparently useful measures will be described.

Barwell, Isaac, and Kunin (2015) examined 16 conceptual properties, and two sampling properties, for 29 measures. They concluded that 'β-diversity is a multifaceted concept' and stated that 'any study measuring β-diversity should be explicit about its goals'. They also observed that no current measures take account of species that are present but unobserved.

10.7.1   Whittaker's β_W

When Whittaker (1960, 1972) introduced the term β-diversity, he suggested that it could be measured by β_W, the extent to which the overall species richness of a region, S, exceeds $\bar{s}$, the average richness of individual localities within that region:

$$\beta_W = \frac{S}{\bar{s}} - 1. \tag{10.23}$$

According to Koleff, Gaston, and Lennon (2003), β_W is much the most frequently used of the possible measures of β-diversity.

The value of β_W is dependent on the sizes of the regions being compared. As a trivial example, suppose that a region consists of four quarters, with a different species (and no other species) present in each quarter. Thus $S = 4$ and, when the quarters are taken as the individual localities, $\bar{s} = 1$ so that $\beta_W = 3$. However, if the individual localities were the two halves of the region, then $\bar{s} = 2$ so that $\beta_W = 1$.

Example 10.7: Bahamian coral

Table 10.1 uses data derived from the online database made available by AGRRA (Atlantic and Gulf Rapid Reef Assessment).[2] The data refer to a particular region of the Andros Shelf in the Bahamas. The table reports the species noted on five transects in 2007 and also two traversed in 2011.

In 2007, a total of 12 species were observed, with an average of 6.4 per transect, giving $\beta_W = 12/6.4 - 1 = 0.88$. In 2011, $S = 10$ and $\bar{s} = 7.5$ so that $\beta_W = 0.33$. Although this suggests a reduction in β-diversity, it would be very unwise to draw any inference from the results for a single site. However, if there was a similar reduction across the majority of the Bahamian region, then this might be a fair conclusion.

Table 10.1 Coral species found using point-intercept sampling (10 equi-spaced points) at site S1–06 on the Andros Shelf in the Bahamas. Five transects were sampled on 9 October 2007 and two on 25 October 2011. All transects had average depths of between 10.7 m and 12.6 m.

Year	Transect	A	B	C	D	E	F	G	H	I	J	K	L	M	Total
2007	1				✓		✓	✓			✓	✓	✓	✓	7
	2				✓	✓					✓	✓		✓	5
	3		✓		✓	✓			✓		✓			✓	6
	4			✓		✓	✓	✓	✓	✓	✓				7
	5			✓		✓	✓	✓	✓		✓	✓			7
2011	1				✓	✓	✓	✓	✓	✓		✓		✓	8
	2	✓		✓	✓	✓		✓	✓					✓	7

Key: A = *Eusmilia fastigiata*; B = *Isophyllia sinuosa*; C = *Millepora alcicornis*; D = *Montastraea cavernosa*; E = *Orbicella annularis*; F = *Orbicella faveolata*; G = *Orbicella franksi*; H = *Porites astreoides*; I = *Porites furcata*; J = *Porites porites*; K = *Siderastrea siderea*; L = *Stephanocoenia intersepta*; M = *Undaria agaricites*.

10.7.2 Beta turnover

Wilson and Shmida (1984), who were studying the variation in species across environmental gradients, suggested that, when comparing two locations, a useful measure would be one indicating the proportion of species present in just one of the two locations. Suppose that there are s_1 species at location 1, with u_1 of these being unique to that location. Define s_2 and u_2 in the corresponding fashion for location 2. Wilson and Schmida's suggestion was to use 'beta turnover' defined by:

$$\beta_T = \frac{u_1 + u_2}{s_1 + s_2}. \tag{10.24}$$

Example 10.8: Bahamian coral (cont.)

For the two 2011 transects, Table 3.1 shows that three of the eight species present in transect 1 (F, I, and K) are not present in transect 2, and that two of the seven species present in transect 2 (A and C) were not observed in transect 1. Thus $\beta_T = (3 + 2)/(8 + 7) = 0.33$.

For the 2007 transects there are ten possible pairs of transects that may be compared. The most similar are transects 4 and 5 ($\beta_T = (1 + 1)/(7 + 7) = 0.14$). The most dissimilar are transects 2 and 4 ($\beta_T = (3 + 5)/(5 + 7) = 0.67$). The average value of β_T over the ten pairs is 0.45, which again appears to suggest that there was a greater β-diversity in 2007 than in 2011.

10.7.3 Beta composition

Lennon et al. (2001) observed that if two locations vary greatly in the number of species present, then there will be many species in the richer location that do not occur in the poorer location. However, if there are any species in the poorer location that do not occur in the richer, then this will certainly indicate a variation in species composition. Using the suffix 1 for the poorer location, they suggested using:

$$\beta_C = \frac{u_1}{s_1}. \tag{10.25}$$

In their comparison of alternative measures, Koleff, Gaston, and Lennon (2003), concluded that, while no measure was entirely satisfactory, β_C was possibly the best. Similarly, Barwell, Isaac, and Kunin (2015) concluded that 'when turnover in rare species is important and undersampling is not severe, the presence-absence metric [β_C] is favoured'.

Example 10.9: Bahamian coral (cont.)

Using the 2011 data in Table 3.1, transect 2 is the poorer location, so that $u_1 = 2$, $s_1 = 7$ and $\beta_C = 2/7 = 0.29$. For the 2007 data, the ten possible pairs give values of β_C varying between 1/7 (for transects 4 and 5) to 3/5 (for transects 2 and 4), with an average of 0.40. Yet again, the conclusion is that there has been a reduction in β-diversity.

10.7.4 Beta overlap

The measures so far discussed, have all been expressed in terms of *numbers of species* present at different locations. Morisita (1959) suggested an index based on the *numbers of individuals* present for each species at each location. Let j and k denote the locations of interest, with n_{ij} being the number of individuals of species i observed at location j, and

$$N_j = \sum_i^S n_{ij}, \tag{10.26}$$

where N_k is similarly defined and S is the total number of different species over the two locations. Let λ_j and λ_k denote the values of Simpson's index (Equation (10.5)) for the two locations. Then Morisita's index is given by

$$\beta_M = 1 - \frac{2}{(\lambda_j + \lambda_k)N_j N_k} \sum_{i=1}^S x_{ij} x_{ik}, \tag{10.27}$$

where λ_j and λ_k are the values of Simpson's index (Section 10.4) for the two locations. In an extreme case, where there was no overlap between the species present at the two locations, the index would have the value 1.

Barwell, Isaac, and Kunin (2015) concluded that β_M 'is the [measure] most independent of sample size [of the 29 measures they considered], at the expense of being almost completely insensitive to turnover in rare species … the emphasis [β_M] places on common species is suitable when shifts in dominance are of interest'.

Example 10.10: Californian trees (cont.)

In this example, the data used are the trees having locations identified in the top and bottom rows of Figure 9.8. Two comparisons will be made: the first compares the information from the cells (labelled α and β in Table 10.2) in the first column for these two rows, while the second uses the more extensive information (labelled γ and δ) provided by the next three columns for those rows. In total there are 12 species involved and the information is summarized in Table 10.2.

Comparing the two single cells (α and β),

$$\sum x_{i\alpha} x_{i\beta} = (4 \times 2) + (11 \times 3) = 41,$$

with

$$\lambda_\alpha = \frac{12 + 110}{16 \times 15} \approx 0.51$$

and $\lambda_\beta = 0.25$. In this case, therefore,

$$\beta_M = 1 - \frac{2 * (8 + 33)}{(0.51 + 0.25) \times 16 \times 8} \approx 0.16.$$

Comparing the rest of the two rows (γ and δ) gives essentially the same result, despite an apparently very different species mix. It is this consistency that makes β_M a reliable measure even with small samples.

Table 10.2 Numbers of each tree species present in the top and bottom rows of that part of the Californian tree plot illustrated in Figure 9.8. Counts are given for four sections labelled α, β, γ, and δ.

	AM	CC	HA	LD	LH	PM	QA	QP	SS	TD	UC	VO	N
					Tree species (see Key)								
α	1	0	0	0	0	4	0	0	11	0	0	0	16
β	0	0	0	0	0	2	0	3	3	0	0	0	8
γ	5	1	3	3	0	7	4	8	9	2	1	0	43
δ	7	10	0	3	1	20	7	11	2	1	0	2	64

Key: α = Top left cell; β = Bottom left cell; γ = Rest of top row; δ = Rest of bottom row; AM = *Arbutus menziesii*; CC = *Corylus cornuta*; HA = *Heteromeles arbutifolia*; LD = *Lithocarpus densiflorus*; LH = *Lonicera hispidula*; PM = *Pseudotsuga menziesii*; QA = *Quercus agrifolia*; QP = *Quercus parvula*; SS = *Sequoia sempivirens*; TD = *Toxicodendron diversilobum*; UC = *Umbellularia californica*; VO = *Vaccinium ovatum*.

10.7.5   Other measures

Harrison, Ross, and Lawton (1992) observed that the value of Whittaker's β_W would be increased by choosing to examine locations with few species. They suggested eliminating this possibility by using $S/\max(s) - 1$ rather than $S/\bar{s} - 1$. For the comparison of the species at two locations, with the notation of Section 10.7.2, their suggestion can be rewritten as

$$\beta_{-1} = \frac{u_1}{s_2}. \tag{10.28}$$

The previous measures in this section have all focused on the differences between the species at a pair of locations. By contrast, *Jaccard's index* addresses their similarities. If c denotes the number of species common to the two locations, then

$$J = \frac{c}{S}, \tag{10.29}$$

where S is the total number of species observed. For comparison with the previous measures one could use $\beta_J = 1 - J$.

Example 10.11: Bahamian coral (cont.)

For the 2011 data, $u_1 = 2$, $s_2 = 8$, $c = 5$, and $S = 10$ giving $\beta_{-1} = 0.25$ and $\beta_J = 0.3$. For 2007, the average values are $\beta_{-1} = 0.35$ and $\beta_J = 0.39$.

In this case, there is very little difference between the values of the various measures. Nevertheless, one should be careful never to compare the value of a measure on one data set, with that of a different measure on a second data set.

11. Species abundance distributions (SADS)

An apparently ubiquitous feature of any mixed population, is that just a few species will account for the majority of individuals, while many species will be represented by just one or two individuals. The underlying mechanisms that lead to this result are not properly understood. McGill et al. (2007), in a masterly review of recent work on this topic, note that 'understanding SADs is a major stepping stone to understanding communities in general'; they list 27 different models that have been proposed. They suggest that this model proliferation results from 'a failure to successfully test and reject theories with data'. In this chapter, only a few of those 27 models will be considered, with the choice being governed by historical importance, frequency of use, and simplicity of description. First, some alternative methods for illustrating abundance data, are presented.

11.1   Illustrating abundance distributions

The difficulty of illustrating an abundance distribution is a direct result of its extreme skewness. In a large sample, the most abundant species may be represented by thousands of individuals, whereas there may be several species known only by the presence of a single individual. To show clearly both ends of the range of abundance will usually imply working with log(abundance) rather than the raw abundance.

11.1.1   Rank-abundance plot

Whittaker (1965) suggested that an effective diagram resulted from plotting abundance (or its logarithm) against the species arranged in decreasing order of abundance to give a rank-abundance plot.

Example 11.1: Indian trees and Californian trees (cont.)

Figure 11.1 compares the tree data using rank-abundance plots. Plot (a) is a simple rank-abundance plot for both sets of tree data, with the proportions accounted for by the ranked species shown on the y-axis. Both sets of data display the characteristic 'hollow-curve' shape. Plot (b) shows the actual abundances, using a log scale. Here the plots have very different slopes because both the sample sizes, and the numbers of species, are very different from one set to the other.

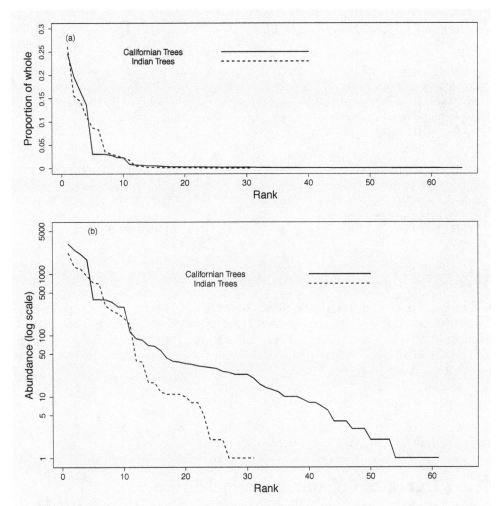

Figure 11.1 (a) Rank-abundance plot for both data sets showing the proportions of the whole set. The two curves are very similar and display the typical 'hollow-curve' shape. (b) As (a) but showing abundances on a log scale.

11.1.2 Octaves

An alternative to using logarithms, is to group the frequencies into ranges and then illustrate the distribution by means of equal-width bars. Preston (1948) suggested using what he termed *octaves*: groups chosen such that the range of each successive group is roughly double that of its predecessor. Gray, Bjørgesæter, and Ugland (2006) examined the merits of minor variants of Preston's original formulation. They recommended using the groupings 1, 2–3, 4–7, 8–15, etc., so that octave m spans the range from 2^m to $2^{m+1} - 1$, with $m = 0, 1, 2, \ldots$. The data are then presented using a grouped bar chart.

Example 11.2: Indian trees (cont.)

Table 11.1 shows the octave frequencies for the Indian tree frequencies (for 2000) listed in the Appendix. there were 12,574 trees in total, so that the most extreme possible octave is number 13, corresponding to a count in the range 8192 to 16,383. An observation in the last category is, of course, extremely unlikely, since it would imply that more than two-thirds of individuals were the same species. These counts are illustrated later in Figure 11.3.

Table 11.1 The octave counts for the Indian tree data for 2000 given in the Appendix.

Octave	0	1	2	3	4	5	6	7	8	9	10	11	12	13
Count	8	7	5	9	10	7	5	0	6	0	1	3	0	0

11.2 The log-series distribution

In a comparison of alternative SADs, Baldridge et al. (2016) examined their fits to more than 16,000 communities taken from a wide range of taxonomic groups and ecosystems. In about 70% of these communities, the lowest AIC values resulted from the use of the log-series distribution. The form of this distribution, which was given in Section 1.5.1, is repeated here for convenience:

$$P_k = \frac{\alpha}{kS} \left(\frac{n}{n+\alpha} \right)^k, \qquad k = 1, 2, 3, \ldots. \qquad (11.1)$$

The procedure for estimating the parameter α was described in Section 10.5.

11.3 Truncated Poisson-lognormal distribution

Having plotted octave-based diagrams for several abundance distributions, Preston (1948) concluded that a lognormal distribution might be appropriate. Bulmer (1974) developed the idea by suggesting using a truncated version[1] of the Poisson-lognormal distribution (Section 1.5.2). The fit of this distribution is easily performed in R:

```
# The species abundances are in the vector counts
library(poilog)            # Calls the appropriate R library
fitpoilog(counts)          # Fits the Poisson-lognormal model
```

The output provides estimates of μ and σ; the fitted probabilities are then obtained using the command:

```
# The probability distribution must be divided by
# (1 - P₀) since zero is impossible
dpoilog(xrange,mu,sigma)/(1-dpoilog(0,mu,sigma))
```

If S is the number of species observed, then an estimate of the number of unobserved species is provided by

Williamson and Gaston (2005) have argued that, despite their popularity, neither the lognormal distribution, nor the Poisson-lognormal distribution is appropriate, They advance several reasons including the observation that these distributions predict 'the existence of many extremely abundant species that do not exist'. On the other hand, in their study of the 35 million observations recorded in the Botanical Information and Ecology Network, Enquist et al. (2019), found that the model providing the lowest AIC value was provided by the Poisson-lognormal distribution.

Example 11.3: Indian trees (cont.)

Figure 11.2 shows a dot plot of the Indian data for 2000, together with the fitted log-series and Poisson-lognormal distributions. The log-series distribution appears to give the better fit at the lowest abundance levels, but the difference between the distributions is not great. For the Poisson-lognormal distribution, the fitpoilog(counts) R function gives estimated values for μ and σ as 2.07 and 2.71, respectively. These values give P_0 = 0.188 implying that the 61 species observed account for just over 80% of the total number: the bottom line is that, using the Poisson-lognormal distribution for these data suggests that there are a further 14 as-yet-unseen species. This is far fewer than the previous estimates.

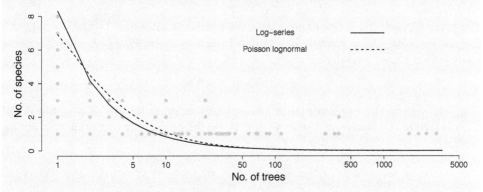

Figure 11.2 Dot plot of the Indian tree data (for 2000), showing the fitted values using the log-series and truncated Poisson-lognormal distributions.

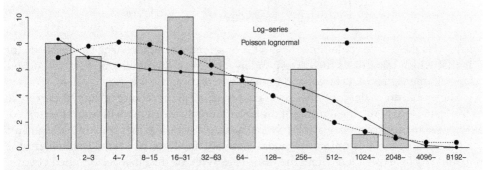

Figure 11.3 Octave diagram of the Indian tree data (for 2000), showing the fitted values using the log-series and truncated Poisson-lognormal distributions.

Figure 11.3 shows the octave plot for the Indian data (see Table 11.1) together with the fitted frequencies for the log-series distribution for which $\alpha = 8.33$ (see Figure 10.4). The final two octaves, with zero observed counts, illustrate the validity of the criticism by Williamson and Gaston (2005) that the right-hand tail of the Poisson-lognormal distribution converges too slowly on zero.

The octave plot was obtained using the `barplot` command from the `gambin` library.

11.4 The gambin model

According to Matthews et al. (2014), when abundance data are summarized using octaves, this model, introduced by Ugland et al. (2007), often provides a superior fit to that of its rivals (though this is not the case fn the example that follows). The name results from the model's use of both the gamma distribution and the binomial distribution.

A random variable X having a gamma distribution with a unit scale parameter, has probability density function f(x) given by

$$f(x) = \frac{1}{\Gamma(\alpha)} x^{\alpha-1} e^{-x}, \qquad\qquad 0 \leq x \leq \infty. \qquad (11.2)$$

The parameter α is the single parameter of the gambin distribution. Ugland et al. (2007) suggest working with a gamma distribution truncated at the 99th percentile. A critical value is then $C_{99}(\alpha)$ given by

$$P(X < C_{99}(\alpha)) = 0.99. \qquad (11.3)$$

The next step in the implementation of the distribution is the division of the range $(0, C_{99}(\alpha))$ into 100 equal-sized steps with P_j being the probability of a value of X falling in step j:

$$P_j = \frac{1}{0.99} P\left(\frac{j-1}{100} C_{99}(\alpha) < X < \frac{j}{100} C_{99}(\alpha)\right), \qquad j = 1, 2, \ldots, 100. \qquad (11.4)$$

With S species observed, the estimated number in octave k is then given by

$$S \sum_{j=1}^{100} P_j \binom{M}{k} \left(\frac{j}{100}\right)^k \left(1 - \frac{j}{100}\right)^{M-k}, \qquad k = 0, 1, 2, \ldots, M, \qquad (11.5)$$

where M is the octave containing the largest observed abundance. The value chosen for α is that which maximizes the correspondence between the fitted and observed octave counts. Large values of α correspond to cases where there are few rare species.

The procedure as described has several arbitrary steps: the choice of the 99th percentile as the truncation point, the limiting of the distribution to the number of octaves observed, as opposed to the maximum feasible (given the sample size), and the choice of 100 segments for summation. For this unimodal modal there are, in effect, two fitted parameters: the maximum octave considered, and the value of α. With the data in a vector called `counts`, the distribution can be fitted and the octaves plotted using the following R commands:

```
library(gambin)
fit1<-fit abundances(counts); plot(fit1)
```

The fit of the model, together with the fitted frequencies and the AIC value, are examined using the commands:

```
fit1; fit1$fitted.values; AIC(fit1)
```

Since many data sets appear to give rise to multimodal octave plots, Matthews et al. (2018) proposed using mixtures of gambins. For example, to fit a bimodal gambin requires:

```
fit2<-fit abundances(counts, no of components=2)
```

For a bimodal gambin model there are five fitted parameters: two for each gambin, together with one describing the balance between the two.

Example 11.4: Indian trees (cont.)

Figure 11.4 shows the fit of the unimodal and bimodal gambin models for the Indian tree data (of 2000). For the bimodal model, the second gambin is truncated at octave 1; this explains the excellent fit to the first two octaves. The octave fits are considered in more detail in the next section.

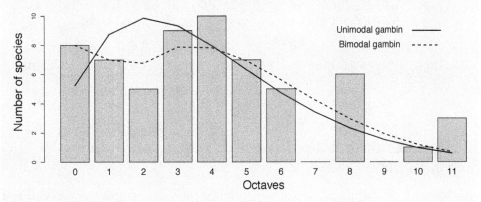

Figure 11.4 Octave plot of the Indian forest trees, showing the fits of the unimodal and bimodal gambin models.

11.5   Testing the goodness of fit of a model to a set of octave counts

Suppose that there is a set of observed octave frequencies $O_0, O_1, ..., O_m$, and a matching set of expected frequencies $E_0, E_1, ..., E_m$ according to some model. The traditional measures of goodness of fit are Pearson's X^2 and the likelihood ratio G^2 (Section 1.8). However, the chi-squared approximations to the distributions of X^2 and G^2 will not be accurate (Upton, 2016) when (as will usually be the case) there are many small expected frequencies. The solution is to combine the counts for adjacent octaves to achieve reasonably large (> 3, say) expected frequencies. If this results in g groups of octaves, then, if the fitted model is correct, both X^2 and G^2 can be regarded as values from a chi-squared distribution with $(g - p - 1)$ degrees of freedom, where p is the number of parameters estimated from the data.

Example 11.5: Indian trees (cont.)

Table 11.2 shows the values for the fits of the log-series, Poisson-lognormal, and the gambin model with either one or two modes. Because of the low expected frequencies for the upper octaves, these have been combined to reduce the number of octaves to ten. Also given in the table are the degrees of freedom for each model, the values of the X^2 goodness of fit statistic, and the tail probability for the corresponding chi-squared distribution. The most plausible fits come from the Poisson-lognormal and log-series distributions. The unimodal gambin appears to give a significantly poor fit.

Table 11.2 The observed octave counts for the Indian tree data given in Table 11.1, together with the fitted values for four models. Also shown are the degrees of freedom, the value of the X^2 goodness of fit statistic, and the tail probability, p.

	Octave										d.f.	X^2	p
	0	1	2	3	4	5	6	7	8	9+			
	8	7	5	9	10	7	5	0	6	4			
LS	8.3	6.9	6.3	6.0	5.8	5.7	5.5	5.1	4.5	6.9	8	12.0	0.15
PL	6.9	7.8	8.1	7.9	7.3	6.3	5.2	4.0	2.9	4.6	7	10.1	0.18
G1	5.3	8.7	9.9	9.3	8.0	6.3	4.7	3.4	2.3	3.1	7	14.4	0.04
G2	8.0	7.0	6.8	7.9	7.8	6.9	5.6	4.2	3.0	3.8	4	8.5	0.07

Key: LS = Log-series; PL = Poisson-lognormal; G1 = Unimodal gambin; G2 = Bimodal gambin

11.6 Determining the drivers for species abundance distributions

Whichever type of distribution is used as a summary of a set of abundance counts, it will have at least one parameter that requires estimation: α for the log-series and unimodal gambin distributions, or μ and σ for Poisson-lognormal distribution. Suppose that there are multiple data sets that are available, all concerned with the same general type of species. In one data set there may be a wide variety of species. In another data set there may be disappointingly few. The species abundance distributions will be very different and so will be the estimated values of the parameters. The question of interest is which characteristics of the locations represented by these data sets are related to the changes in parameter values. One approach (Yen, Thomson and Mac Nally, 2012) is to first obtain parameter estimates for each fitted abundance distribution, and then to use these as the values for a response variable dependent on, for example, a regression model involving one or more explanatory variables. A more elegant approach is to use *functional data analysis* in which the entire collection of data sets are treated simultaneously, with their unknown parameters linked, as required, to the underlying explanatory variables. This approach is discussed by Yen et al. (2014), and by Matthews et al. (2017).

12. Other aspects of diversity

12.1 Evenness

Evenness (also called *equitability*) is intended to be a quantification of the extent to which the species present are equally abundant. Evenness was described by Magurran, Queiroz and Hercos (2013) as 'a key measure of community structure'. The most frequently used measure of evenness is probably that suggested by Pielou (1966):

$$E_P = \frac{H}{\max(H)} = \frac{H}{\ln(S)},$$

(12.1)

where H is Shannon entropy (Section 10.2). The quantity E_P takes values between 0 and 1. However, just as changes in the value of H are difficult to interpret, so are changes in E_P (Alatalo, 1981).

Hill (1973) suggested that a ratio of any two effective numbers (Section 10.4) would provide a measure of evenness, since the ratio equals 1 only in the case of equally likely species. Sheldon (1969) had used

$$E_S = \frac{\exp(H)}{S}$$

(12.2)

as a measure of *evenness*; this index satisfies Hill's suggestion, since $S = N_0$ and $\exp(H) = N_1$. However, as Heip (1974) noted, E_S has a minimum value of $1/S$ rather than zero. Heip therefore suggested using

$$E_H = \frac{\exp(H) - 1}{S - 1},$$

(12.3)

since this has a range (from 0 to 1) that does not depend on S.

Kvålseth (2015) examined the properties of 11 measures of evenness, including E_P and E_H. He set out five desirable properties and demonstrated that few measures possessed all of them. One measure that does, which Kvålseth termed the 'gold standard', was that suggested by Williams (1977):

$$E_W = 1 - \sqrt{\frac{S \sum_{i=1}^{S} p_i^2 - 1}{S - 1}},$$

(12.4)

where p_i is the proportion of individuals belonging to species i. E_W is linearly related to the standard deviation of the p_i-values.

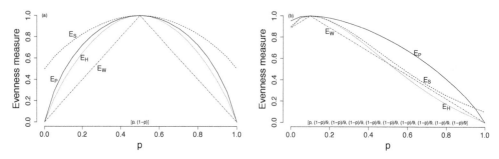

Figure 12.1 Comparison of the four evenness functions E_P, E_S, E_H, and E_W, as p varies. (a) Two species with proportions p and $(1 − p)$. (b) Ten species, one with proportion p and nine with proportions $(1 − p)/9$.

Figure 12.1 shows the behaviour of the four measures introduced here for two cases. In the first case there are just two species with probabilities p and $(1 − p)$. All the indices equal 1 when $p = 0.5$, but the values of E_S are the greatest.

In the second case there are ten species, with one species accounting for a proportion p and the remaining nine being equally likely. It is reassuring that, in this case, with the exception of the generally inflated values for E_P, the values of the alternative measures do not greatly differ from one another. In both cases E_W is linearly dependent on the largest proportion.

12.2 Similarity and complementarity

A question of possible interest, is the extent to which one collection of species resembles another. Let S_i denote the number of species in collection i ($i = 1$, 2) and let S_{12} be the number of species occurring in both collections. Many measures of similarity have been proposed, of which the most familiar is probably *Jaccard's index* given by:

$$J = \frac{S_{12}}{S_1 + S_2 − S_{12}}. \tag{12.5}$$

The obvious interpretation of J is as the proportion of observed species that were present in *both* collections. *Complementarity* may be similarly defined as the proportion of observed species that were present in *just one* of the collections:

$$C = 1 − J = \frac{S_1 + S_2 − 2S_{12}}{S_1 + S_2 − S_{12}}. \tag{12.6}$$

Chao et al. (2006) demonstrate that J may also be interpreted as the probability that, when a species is chosen at random from one collection, and a species is chosen at random from the other collection, both the chosen species are present in both collections.

Chao et al. (2006) then extend that idea to take account of the numbers of individuals. Instead of working with S_1 and S_2 the formula now uses U and V, where U is the proportion of the individuals in the first collection that belong to species present in the second collection, with V being the proportion of the individuals in the second collection that belong to species present in the first collection. The equivalent of J is then J^* given by

$$J^* = \frac{UV}{U + V − UV}. \tag{12.7}$$

Chao et al. (2006) refer to J^* as the abundance-based unadjusted Jaccard index. To take account of unseen species, they propose an adjustment based on species that occur in both collections, but are scarce in at least one collection. The required notation is as follows:

f_{i+} the numbers of species that occur in both collections, but are represented by just i ($i = 1, 2$) individuals in the first collection.

f_{+i} the numbers of species that occur in both collections, but are represented by just i ($i = 1, 2$) individuals in the second collection.

N_j the total number of individuals in collection j ($j = 1, 2$)

n_j the number of species represented by just one individual in collection j ($j = 1, 2$)

With these definitions, their revised version of the Jaccard index is:

$$J' = \frac{U'V'}{U' + V' - U'V'},$$ (12.8)

where

$$U' = U + \frac{N_2 - 1}{2N_1 N_2} \frac{f_{+1}}{\max(1, f_{+2})} n_1,$$

$$V' = V + \frac{N_1 - 1}{2N_1 N_2} \frac{f_{1+}}{\max(1, f_{2+})} n_2.$$

The idea of the adjustment is to take account of species that are present in the population, but do not occur in one (or both) of the two collections. As a consequence $J' > J^*$ (while also $J^* > J$).

Example 12.1: Costa Rican ants (cont.)

The ants in Costa Rica were trapped in various ways[1] with $S_1 = 165$ species ($N_1 = 3262$ individuals) being found using fogging, and $S_2 = 197$ species ($N_2 = 1415$ individuals) being found using the Winkler sifter. There were $S_{12} = 38$ species that were trapped by both methods. Thus

$$J = \frac{38}{165 + 197 - 38} = 0.12.$$

The 38 species observed by both methods included 870 individuals observed using fogging and 385 observed using the Winkler sifter, so that

$$U = \frac{870}{3262} = 0.2667, \quad V = \frac{385}{1415} = 0.2721,$$

and hence

$$J^* = \frac{0.2667 \times 0.2721}{0.2667 + 0.2721 - (0.2667 \times 0.2721)} = 0.16.$$

Among the 38 shared species, there were $f_{1+} = 11$ species represented by one individual in the fogging data. The frequencies for these species in the second collection were 18, 3, 11, 21, 1, 26, 29, 18, 3, 5, and 4, so that $n_2 = 139$. The other relevant values are $f_{2+} = 5$, $f_{+1} = 10$, $f_{+2} = 2$, and $n_1 = 219$. Adjusting for unobserved species gives: $U' = 0.4344$, $V' = 0.3801$, and hence

$$J' = \frac{0.4344 \times 0.3801}{0.4344 + 0.3801 - (0.4344 \times 0.3801)} = 0.25.$$

The final value is roughly double that given by the Jaccard index, which takes no account of the numbers of individuals.

12.3 Turnover

Yuan et al. (2016) suggested that a comparison of the species mixture at one location with that at a nearby location within the same region would provide a measure of diversity. They refer to this as *spatial turnover* and suggested several possible measures. A second type of turnover is *temporal turnover* which refers to changes in species composition at a single location over time. A simple and effective measure uses the so-called *Manhattan distance* given by:

$$d_M = \frac{1}{2} \sum_{k=1}^{K} |p_{1k} - p_{2k}|, \tag{12.9}$$

where p_{ik} (which may be zero) is the proportion of the total count at time i ($i = 1, 2$) which consists of species k. The bounds on d_M are 0 (no change) and 1 (complete change).

Example 12.2: Indian trees (cont.)

The Appendix lists the trees counted in the Mudumalai Forest Dynamics Plot in 1988, 1992, and 2000. The values of d_M for the changes from 1988 to 1992, from 1992 to 2000, and from 1998 to 2000, are 5.3%, 7.0%, and 12.1%, respectively.

12.4 Rarity

In a fascinating and influential paper, Rabinowitz (1981) suggested that there were three relevant aspects to be considered in determining whether a species should be regarded as rare. These were:

- Geographic range – large or small;
- Habitat specificity – wide or narrow;
- Local population size – large and occasionally dominant, or small and never dominant.

Cross-classifying these aspects gives eight combinations, with seven of them describing some aspect of rarity; the rarest being a small population occupying a narrow habitat in a restricted location. Yu and Dobson (2001), who considered more than a thousand species of mammals, categorized more than one-third as belonging to this rarest category. Similarly, Espeland and Emam (2011), using published studies of plants, found that 30% of the plants studied fell in the rarest category.

Location in Britain can be specified by reference to a grid of 1 km squares. A simple method for quantifying geographic range is therefore provided by counting the number

Table 12.1 The categorization used by Rabinowitz (1981) to determine the rarity of a species, with the values of the vulnerability index used by Fattorini et al. (2012).

Range: (W, Widespread; R, Restricted)	W	W	W	R	W	R	R	R
Habitat: (V, Varied; S, Specific)	V	V	S	V	S	V	S	S
Population: (A, Abundant; L, Low)	A	L	A	A	L	L	A	L
Vulnerability index, a	1	2	3	4	5	6	7	8

of grid squares within which the species of interest has been recorded. However, Hartley and Kunin (2003) demonstrated that the relative abundance of two species can be critically dependent on the size of square used. They also noted that the number of occupied squares is dependent on the precise placement of the grid.

Using Rabinowitz's categories, Fattorini et al. (2012) assigned a vulnerability index (a; see Table 12.1) to each species in order to identify locations of prime conservation concern. They determined two quantities:

$$\beta_C = \frac{\sum_{i=1}^{L}(\alpha_i - \alpha_{\min})}{L(\alpha_{\max} - \alpha_{\min})}, \tag{12.10}$$

$$\beta_W = \frac{\sum_{i=1}^{L}(\alpha_i - \alpha_{\min})}{\sum_{i=1}^{S}(\alpha_i - \alpha_{\min})}, \tag{12.11}$$

where L is the number of species at the local site under consideration, S is the total number of species across all sites, α_i is the vulnerability score for species i (see Table 12.1), and $\alpha_{\max}$ and $\alpha_{\min}$ were the maximum and minimum values across all species. The β_C index helps identify sites with a high proportion of threatened species, while β_W identifies sites with large numbers of threatened species.

Example 12.3: Californian trees (cont.)

The Californian study region contained 8372 trees representing 31 tree species. As usual the counts presented a highly skewed distribution. In this case there were five species represented by a single tree, and one species (*Pseudotsuga menziesii*) accounting for 2180 trees. A species was defined to be abundant if there were more than the median number of individuals.

The study region was divided into 600 square quadrats with side 10 m, with the number of quadrats containing a species being recorded for each species. There were five species recorded in more than 200 of the quadrats, with *Pseudotsuga menziesii* being the most widespread (293 quadrats). A species was regarded as widespread if it occurred in more than the median number of quadrats occupied by a species, which was 10.

In the absence of information on habitat, a species was regarded as having specific needs if the quadrat counts for that species had a coefficient of variation (see Section 1.2.6) that was greater than the median for the 31 species.

With these definitions, 14 of the 31 species were assigned an α-value of 1. One species (*Pinus ponderosa*) was assigned the value 3, one species (*Umbellularia californica*) was given the value 6, and the remainder were given the value 8.

Figure 12.2 represents the β_C values for the 600 quadrats, with the darkest squares having the highest values. There were 518 quadrats with zero values: each of the trees present in these quadrats was judged to have $\alpha = 1$. The numbers of vulnerable species (using the criteria outlined) are shown in the figure for each quadrat. A comparison of local characteristics (soil, rainfall, etc.) might help to identify the conditions required for the scarcer species to flourish. Varying the criteria used for each characteristic might make any such relationship more evident.

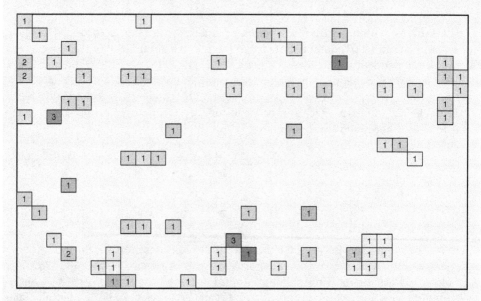

Figure 12.2 The variation in β_C values across the 300 m × 200 m Californian plot. Higher values are represented by darker squares. The number of more vulnerable species is indicated for each quadrat.

Appendix

The table lists species of trees in the 50-ha (500 m x 1 km) Mudumalai Forest Dynamics Plot, which forms part of the Mudumalai Wildlife Sanctuary in the Western Ghats of India. The trees listed have diameters of at least 10 cm at breast height. With the notable exception of Indian laburnum (*Cassia fistula*), there has been a decline in the numbers of most species, while the number of species has also declined from 63 in 1988 and 1992 to 61 in 2000. The most frequent are axlewood (*Anogeissus latifolia*), laurel (*Terminalia crenulata*), myrtle (*Lagerstroemia microcarpa*), and teak (*Tectona grandis*).

Order	Family	Genus	Species	1988	1992	2000
B	Boraginaceae	Cordia	obliqua	18	20	23
B	Boraginaceae	Cordia	wallichii	37	38	38
C	Celastraceae	Cassine	glauca	4	3	1
E	Ebenaceae	Diospyros	montana	130	117	114
E	Lecythidaceae	Careya	arborea	28	25	20
E	Sapotaceae	Madhuca	neriifolia	3	3	2
F	Fabaceae	Albizia	odoratissima	4	3	2
F	Fabaceae	Bauhinia	malabarica	27	26	25
F	Fabaceae	Bauhinia	racemosa	9	9	7
F	Fabaceae	Cassia	fistula	179	246	289
F	Fabaceae	Dalbergia	latifolia	33	35	30
F	Fabaceae	Erythrina	indica	5	5	3
F	Fabaceae	Ougeinia	oojeinensis	109	94	57
F	Fabaceae	Pterocarpus	marsupium	17	17	12
F	Faboideae	Butea	monosperma	32	31	29
G	Apocynaceae	Wrightia	tinctoria	1	1	3
G	Rubiaceae	Canthium	dicoccum	19	16	16
G	Rubiaceae	Hymenodictyon	orixense	10	10	10
G	Rubiaceae	Xeromphis	spinosa	558	521	387
G	Rubiaceae	Mitragyna	parvifolia	15	14	14
L	Bignoniaceae	Radermachera	xylocarpa	309	308	293
L	Bignoniaceae	Stereospermum	colais	89	90	89
L	Lamiaceae	Gmelina	arborea	58	51	34
L	Lamiaceae	Premna	tomentosa	2	2	1
L	Lamiaceae	Tectona	grandis	1805	1779	1716
L	Oleaceae	Olea	dioica	4	4	4
L	Oleaceae	Schrebera	swietenioides	69	57	32

Order	Family	Genus	Species	1988	1992	2000
L	Lamiaceae	Vitex	altissima	1	1	1
Mp	Euphorbiaceae	Antidesma	diandrum	1	1	0
Mp	Euphorbiaceae	Mallotus	philippensis	8	9	4
Mp	Phyllanthaceae	Bischofia	javanica	1	1	0
Mp	Phyllanthaceae	Bridelia	retusa	40	30	23
Mp	Phyllanthaceae	Emblica	officinalis	518	464	387
Mp	Salicaceae	Casearia	esculenta	39	40	37
Mp	Salicaceae	Flacourtia	indica	7	7	2
Mv	Bombacaceae	Bombax	ceiba	35	35	35
Mv	Dipterocarpaceae	Shorea	roxburghii	6	6	1
Mv	Malvaceae	Eriolaena	quinquelocularis	242	184	26
Mv	Malvaceae	Grewia	tiliifolia	493	432	353
Mv	Malvaceae	Helicteres	isora	7	8	2
Mv	Malvaceae	Kydia	calycina	1328	635	84
My	Combretaceae	Anogeissus	latifolia	2180	2158	2103
My	Combretaceae	Terminalia	bellirica	34	33	31
My	Combretaceae	Terminalia	chebula	59	51	43
My	Combretaceae	Terminalia	crenulata	2628	2584	2482
My	Lythraceae	Lagerstroemia	microcarpa	3160	3189	3092
My	Lythraceae	Lagerstroemia	parviflora	82	75	67
My	Myrtaceae	Syzygium	cumini	398	393	383
R	Moraceae	Artocarpus	gomezianus	1	1	1
R	Moraceae	Ficus	benghalensis	3	3	3
R	Moraceae	Ficus	drupacea	4	4	4
R	Moraceae	Ficus	hirsuta	1	0	0
R	Moraceae	Ficus	religiosa	7	6	8
R	Moraceae	Ficus	tsjakela	10	10	10
R	Moraceae	Ficus	virens	11	13	10
R	Rhamnaceae	Ziziphus	rugosa	8	7	8
R	Rhamnaceae	Ziziphus	xylopyrus	20	14	6
S	Anacardiaceae	Lannea	coromandelica	13	13	13
S	Anacardiaceae	Mangifera	indica	4	4	1
S	Anacardiaceae	Semecarpus	anacardium	12	12	9
S	Burseraceae	Garuga	pinnata	29	26	23
S	Meliaceae	Chukrasia	tabularis	1	1	1
S	Sapindaceae	Allophylus	cobbe	0	2	1
S	Sapindaceae	Schleichera	oleosa	72	69	69
TOTAL				15,037	14,046	12,574

Key: B = Boriginales; C = Celastrales; E = Ericales; F = Fabales; G = Gentianales; L = Lamiales;
Mp = Malpighiales; Mv = Malvales; My = Myrtales; R = Rosales; S = Sapindales
Source: Center for Tropical Forest Science, Smithsonian Tropical Research Institute

Notes

Chapter 1: Statistical ideas

1. Details taken from the online report by Jhala, Y. V., Qureshi, Q., and Nayak, A. K. (eds). (2019) *Status of tigers, co-predators and prey in India 2018. Summary Report*. National Tiger Conservation Authority, Government of India, New Delhi & Wildlife Institute of India, Dehradun. TR No./2019/05.

2. The data were obtained via the website https://forestgeo.si.edu/sites/north-america/university-california-santa-cruz.

3. For a more precise 95% the value is 1.96σ, but, for the measurement of abundance (given other uncertainties), using 2σ should give appropriate accuracy.

4. Here k is a whole number with a value obtained using the given expression. However, k factorial exists for *all* non-integer positive k. Using the R programming language its value is obtained by writing `factorial(x)`.

5. The value of 1.92 is one-half of the upper 5% value of a chi-squared distribution with 1 degree of freedom. For a 99% interval replace 1.92 by 3.32.

6. Here 3.00 is an approximation to one-half of 5.991, where 5.991 is the upper 5% point of a chi-squared distribution. For the 99% region replace 3.00 by 4.61.

7. The natural logarithm may also be written $\log_e$ since it expresses numbers as powers of e. Thus $\ln(e) = 1$.

8. Using R, the appropriate command is `1 - pchisq(10.86,2)`.

9. In practice the values used are $1/2n, 3/2n, \ldots, (1 - 1/2n)$.

Chapter 2: Quadrats and transects

1. In this chapter it is assumed that the entire contents of the quadrat or transect are observed. Buckland et al. (2001) use the term *strip transect* to distinguish this case from the case where the probability of an item being observed depends upon that item's distance from the observer. For that case, which is considered in Chapter 6, Buckland et al. (2001) used the term *line transect*.

2. For a 25×25 study region, with 25 quadrats, this type of design is easily achieved as follows. The 25 x co-ordinates are 1, 2, ..., 25, while the corresponding y co-ordinates are a random reordering (e.g. using balls drawn from a bag, or a computer's random number generator) of 1, 2, ..., 25.

3. From website https://daac.ornl.gov/cgi-bin/dsviewer.pl?ds_id=1365. The site is Colville201005.

4. http://www.agrra.org/data-explorer/.

5. Such quantities could arise if a 10×10 grid were superimposed on a photograph of the study region.

6. A formal test, assuming a normal (Gaussian) distribution for the individual estimates, would compare the ratio with the upper tail of an F-distribution having $(Q - 1)$ and $Q(n - 1)$ degrees of freedom.

7. Note that the table gives more decimal places than may seem sensible. The reason is that, in calculations, premature rounding may lead to wildly inaccurate final results. Fortunately, computers do not round prematurely!

Chapter 3: Points and lines

1. A consequence of adding constants to the numerator and denominator is that estimates of 0 or 1 (presumed to be impossible values) are avoided.

Chapter 4: Distance methods

1. Here 'plants' is used as an all-embracing term for the individuals being counted, which might be stationary items of any type.
2. The study area is taken to be the unit square. For the random pattern, each plant location is defined by taking a pair of uniform random numbers as the plant co-ordinates. For the gradient pattern a third random number is generated. If this exceeds the second number, then all three numbers are discarded and a new set is considered.

   For the regular pattern, potential plant positions were again generated using a pair of random numbers, but the pair were only accepted if every previously generated location was at least a distance of 0.025 away from the new position.

   The first step in generating a clustered pattern was to randomly locate a cluster centre in an inner square with sides at 0.1 and 0.9. A random integer between 1 and 20 determined the number of individuals in this cluster. The location of each individual was then random within a square of side 0.2 centred on the cluster centre.
3. Quotations are from the translation of the original Japanese.
4. Data kindly supplied by Dr Hijbeek.
5. The values used for k are 3 (the largest value used in many comparative studies), 4 (to match the 4 distances used by the point-centred quarter methods discussed later), and 6, the value recommended by Prodan (1968). For Sites 1 and 2, the 25 sampling points were arranged centrally in a single row at intervals of 4 m (Set 1) or 3 m (Set 2). For Sites 3 and 4, a 5 × 5 lattice of sampling points was used, centrally placed in each site, with 4 m between successive points in a row or column.
6. Diggle used n rather than $1.2(n - 1)$, but the simulations suggest that the latter is appropriate.
7. Byth used 0.25 rather than 0.3 (omitting the 1.2 scaling).

Chapter 5: Variable sized plots

1. Tree diameters are measured at 'breast height' which is typically taken to be 4.5 ft (1.37 m) above ground level. This is familiarly referred to as DBH.

Chapter 6: Quadrats, transects, points, and lines – revisited

1. To obtain estimates of abundance, Mathews et al. (2018) estimated bat densities by 'multiplying the typical maternity roost density in an average quality landscape by twice the typical number of adult females per roost'.
2. Lovers of Wood Thrushes should note that the characteristics of the sites as given here are fictitious.
3. For comparison, a 120° sector with radius 15 m would have an area of about 240 m².
4. Note that the choice of 3 km² is arbitrary, being based on an inspection of Figure 6.3.
5. The under-estimate was attributed to unhelpful weather conditions.
6. $2 \times (5.63 + 2.05) = 15.36$.

Chapter 7: Capture-recapture methods

1. Although Jolly and Cormack had rooms in the same corridor in Aberdeen University (and played each other at chess) they were unaware of each other's research until their papers were published.
2. The corresponding 99% interval is obtained by replacing –1.6 by –2.2 and 2.4 by 3.0 in Equation (7.25).
3. BUGS code may be run under R using the *rjags* package.

Chapter 8: Distance methods

1. The first two Hermite polynomials used in the *Distance* programme are $H_4(z) = 16z^4 - 48z^2 + 12$ and $H_6(z) = 64z^6 - 480z^4 + 720z^2 - 120$.
2. Their further suggestion is to ignore observations for which $p(d) < 0.10$ for point transects, or < 0.15 for line transects.
3. I am particularly grateful to Dr Eric Rexstad for his patient assistance with my use of *Distance*.
4. The probability density function is simply the product of the distance d and the detection function $p(d)$.
5. In 2003 the corresponding figure was much higher at about 49.5 per ha.
6. *Distance* does not permit simultaneous use of covariates and extensions and does not allow a uniform key with covariates.
7. In *Distance* these cases require NAs for the distance entry.
8. These are the numbers after truncation of the distance data.

Chapter 9: Species richness

1. Details are given in the Appendix.
2. In reality, because of spatial clustering, the required number of trees would be greater, but probably no more than a thousand.
3. This is the estimator introduced with different notation in Equation (7.14) in the discussion of capture-recapture methods.
4. For package details, downloads, etc., see http://chao.stat.nthu.edu.tw/wordpress/software_download/.
5. The ant data are freely available at http://esapubs.org/archive/ecol/E083/011/suppl-1.htm.

Chapter 10: Diversity

1. The solution is to work with $\exp(H)$ rather than H itself.
2. Data are available via http://www.agrra.org/data-explorer/.

Chapter 11: Species abundance distributions (SADS)

1. Truncated because a count of zero cannot occur.

Chapter 12: Other aspects of diversity

1. Details of the methods used are given in the appendix to Longino, Coddington, and Colwell (2002).

Further reading

Wide-ranging books

Henderson, P. A., and Southwood, T. R. E. (2016) *Ecological Methods*, 4th edn. Chichester: Wiley. Now in its 4th edition, this is 600 pages of useful information.

Krebs, C. J. (1998) *Ecological Methodology*, 2nd edn. Pearson: Cambridge. The 2nd edition is still available, while some chapters of a possible 3rd edition may be downloaded from the author's website (http://www.zoology.ubc.ca/~krebs/books.html).

Sutherland, W. L. (ed.) (2006) *Ecological Census Techniques*, 2nd edn. Cambridge: Cambridge University Press. In addition to discussing methods from most of the chapters in this present book, this authoritative compilation presents separate chapters relating to amphibians, birds, fish, invertebrates, mammals, plants, and reptiles. Each chapter is written by specialists in their field.

Chapter 1: Statistical ideas

There are many introductory statistics books, and many all-embracing ecology books will include a statistical introduction.

Schreuder, H. T., Ernst, R., and Ramirez-Maldonado, H. (2004) *Statistical Techniques for Sampling and Monitoring Natural Resources*, General Technical Report RMRS-GTR-126. Rocky Mountain Research Station, Fort Collins, CO: US Department of Agriculture, Forest Service. A useful example, available online, and aimed especially at foresters.

Chapter 3: Points and lines

Gregoire, T. G., and Valentine, H. T. (2007) *Sampling Strategies for Natural Resources and the Environment*. Boca Raton, FL: Taylor & Francis. An authoritative text on sampling, with emphasis on sampling in forestry.

Herrick, J. E., Van Zee, J. W., McCord, S. E., Courtright, E. M., Karl, J. W., and Burkett, L. M. (2017) *Monitoring Manual for Grassland, Shrubland, and Savanna Ecosystems*. Las Cruces, NM: USDA-ARS Jornada Experimental Range. Available online, this describes the practicalities of setting up transects on grassland.

Hill, J., and Wilkinson, C. (2004) *Methods for Ecological Monitoring of Coral Reefs*. Townsville: Australian Institute of Marine Science. Available online, this provides very detailed guidance on all aspects of monitoring coral reefs. It includes descriptions of ten applications of transects.

Lutes, D. C., Keane, R. E., Caratti, J. F., Key, C. H., Benson, N. C., Sutherland, S., and Gangi, L. J. (2006) *FIREMON: Fire Effects Monitoring and Inventory System*. Rocky Mountain Research Station, Fort Collins, CO: US Department of Agriculture. Available online, this provides a comprehensive account of the implementation of the line-intercept and point-intercept methods.

Chapter 6: Quadrats, transects, points, and lines – revisited

Gibbs, J. P. (2000) Monitoring populations. In Boitani, L., and Fuller, T. K. (eds), *Research Techniques in Animal Ecology*. New York: Columbia University Press. ch. 7. Available online, this provides a discussion of the use of indices.

Kéry, M., and Royle, J. A. (2015) *Applied Hierarchical Modeling in Ecology: Analysis of Distribution, Abundance and Species Richness in R and BUGS: Volume 1: Prelude and Static Models*; (2020) *Volume 2: Dynamic and Advanced Models*. London: Academic Press. More suitable for statisticians, these two volumes (well over 1000 pages) provide the definitive guide to hierarchical models, including use of the *unmarked* package.

Ralph, C. J., Sauer, J. R., and Droege, S. (1995) *Monitoring Bird Populations by Point Counts*. Albany, CA: US Forest Service. Freely available online.

The European Bird Census Council has provided a useful summary of methods used in Europe which is available at https://pecbms.info/best-practice-guide/.

Chapter 7: Capture-recapture methods

Otis, D. L., Burnham, K. P., White, G. C., and Anderson, D. R. (1978) Statistical inference from capture data on closed animal populations, *Wildlife Monographs*, **62**, 3–135. The classic survey of types of capture-recapture model in closed populations.

The *MARK* manual (with more than 1000 pages) freely available from http://www.phidot.org/software/mark/docs/book/ is probably the most useful guide for any novice wishing to analyse capture-recapture data. The length is an indication of the effort required for a full understanding, and an indication also of the extent to which this chapter has done no more than graze the surface of the topic.

For statisticians, some specialist books are:

McCrea, R. S., and Morgan, B. J. T. (2014) *Analysis of Capture-Recapture Data*. Boca Raton, FL: Chapman and Hall/CRC.

Royle, J. A., Chandler, R. B., Sollmann, R., and Gardner, B. (2013) *Spatial Capture-Recapture*. Waltham, MA: Academic Press.

Amstrup, S. C., McDonald, T. L., and Manly, B. F. J. (2005) *Handbook of Capture-Recapture Analysis*. Princeton, NJ: Princeton University Press.

Two books recommended by the authors of the *MARK* manual are:

Burnham, K. P., and Anderson, D. R. (2002) *Model Selection and Multimodel Inference: A Practical Information-Theoretic Approach*, 2nd edn. New York: Springer-Verlag.

Williams, B. K., Nichols, J. D., and Conroy, M. J. (2002) *Analysis and Management of Animal Populations*. San Diego, CA: Academic Press.

Chapter 8: Distance methods

While the online webpages for the *Distance* programme are a treasure trove of worked examples and videos, with a free online course, the book that underlies the *Distance* approach is:

Buckland, S. T., Anderson, D. R., Burnham, K. P., Laake, J. L., Borchers, D. L., and Thomas, L. (2001) *Introduction to Distance Sampling*. Oxford: Oxford University Press.

The same authors edited a follow-up book that addresses various specialist topics:

Buckland, S. T., Anderson, D. R., Burnham, K. P., Laake, J. L., Borchers, D. L., and Thomas, L. (2004) *Advanced Distance Sampling*. Oxford: Oxford University Press.

An important paper that suggests strategies for obtaining more accurate estimates is given by:
Buckland, S. T., Marsden, S. J., and Green, R. E. (2008) Estimating bird abundance: making methods work, *Bird Conservation International*, **18**, S91–S108.

Part IV: Species

Kindt, R., and Coe, R. (2005) *Tree Diversity Analysis: A Manual and Software for Common Statistical Methods for Ecological and Biodiversity Studies*. Nairobi: World Agroforestry Centre (ICRAF). Available online, a book that concentrates on the measurement of species richness and diversity in the context of trees, and also includes some advanced statistical techniques.

Magurran, A. E., and McGill, B. J. (eds) (2010) *Biological Diversity: Frontiers in Measurement and Assessment*. Oxford: Oxford University Press. Twenty chapters dealing with diversity, SADs, and a variety of applications.

Kondratyeva, A., Grandcolas, P., and Pavoine, S. (2019) Reconciling the concepts and measures of diversity, rarity and originality in ecology and evolution, *Biological Reviews*, **94**, 1317–1337. A detailed and thoughtful paper with contents that match its title.

References

Affleck, D. L. R., Gregoire, T. G., and Valentine, H. T. (2005) Design unbiased estimation in line intersect sampling using segmented transects, *Environmental and Ecological Statistics*, **12**, 139–154.

Agresti, A., and Couli, B. A. (1998) Approximate is better than exact for interval estimation of binomial parameters. *American Statistician*, **52**, 119–126.

Alatalo, R. V. (1981) Problems in the measurement of evenness in ecology, *Oikos*, **37**, 199–204.

Alldredge, M. W., Pollock, K. H., Simons, T. R., Collazo, J. A., and Shriner, S. A. (2007) Time-of-detection method for estimating abundance from point-count surveys, *The Auk*, **124**, 653–664.

Anderson, M. J., Crist, T. O., Chase, J. M., Vellend, M., Inouye, B. D., Freestone, A. L., Sanders, N. J., Cornell, H. V., Comita, L. S., Davies, K. F., Harrison, S. P., Kraft, N. J. B., Stegen, J. C., and Swenson, N. G. (2011) Navigating the multiple meanings of β-diversity: a roadmap for the practicing ecologist, *Ecology Letters*, **14**, 19–28.

Arnason, A. N., Schwarz, C. J., and Gerrard J. M. (1991) Estimating closed population size and number of marked animals from sighting data, *Journal of Wildlife Management*, **55**, 718–730.

Arrhenius, O. (1921) Species and area, *Journal of Ecology*, **9**, 95–99.

Baillargeon, S., and Rivest, L.-P. (2007) Rcapture: loglinear models for capture-recapture in R, *Journal of Statistical Software*, **19**, 1–31.

Baldridge, E., Harris, D. J., Xiao, X., and White, E. P. (2016) An extensive comparison of species-abundance distribution models, *PeerJ*, **4**, e2823.

Barkman, J. J., Doing, H., and Segal, S. (1964) Kritische bemerkungen und vorschläge zur quantitativen vegetationsanalyse, *Acta Botanica Neerlandica*, **13**, 394–419.

Bart, J., and Earnst, S. (2002) Double sampling to estimate density and population trends in birds, *The Auk*, **119**, 36–45.

Bart, J., Droege, S., Geissler, P., Peterjohn, B., and Ralph, C. J. (2004) Density estimation in wildlife surveys, *Wildlife Society Bulletin*, **32**, 1242–1247.

Barwell, L. J., Isaac, N. J. B., and Kunin, W. E. (2015) Measuring β-diversity with species abundance data, *Journal of Animal Ecology*, **84**, 1112–1122.

Beenaerts, N., and Vanden Berghe, E. (2005) Comparative study of three transect methods to assess coral cover, richness and diversity, *Indian Journal of Geo-Marine Sciences*, **4**, 29–37.

Berger, W. H., and Parker, F. L. (1970) Diversity of planktonic foraminifera in deep sea sediments, *Science*, **168**, 1345–1347.

Besag, J. E., and Gleaves, J. T. (1973) On the detection of spatial pattern in plant communities, *Bulletin of the International Statistical Institute*, **45**, 153–158.

Bibby, C. J., Burgess, N. D., Hill, D. A., and Mustoe, S. (2000) *Bird Census Techniques*. London: Academic Press.

Bitterlich, W. (1948) Die Winkelzählprobe. *Allgemeine Forst- und Holzwirtschaftliche Zeitung*, **59**, 45.

Böhning, D. (2008) A simple variance formula for population size estimators by conditioning, *Statistical Methodology*, **5**, 410–423.

Bonar, A. A., Fehmi, J. S., and Mercado-Silva, N. (2011) An overview of sampling issues in species diversity and abundance surveys. In A. E. Magurran and B. J. McGill (eds), *Biological Diversity*. Oxford: Oxford University Press. pp. 11–24.

Bonham, C. D. (2013) *Measurements for Terrestrial Vegetation*, 2nd edn. New York: Wiley.

Boose, E. R., Boose, E. F., and Lezberg, A. L. (1998) A practical method for mapping trees using distance measurements, *Ecology*, **79**, 819–827.

Borchers, D. L., and Efford, M. G. (2008) Spatially explicit maximum likelihood methods for capture-recapture studies, *Biometrics*, **64**, 377–385.

Borchers, D. L., and Fewster, R. M. (2016) Spatial capture-recapture models, *Statistical Science*, **31**, 219–232.

Botta-Dukát, Z. (2005) Rao's quadratic entropy as a measure of functional diversity based on multiple traits, *Journal of Vegetation Science*, **16**, 533–540.

Braun-Blanquet, J. (1928) *Pflanzensoziologie: Gründzuge der Vegetationskunde.* Berlin: Springer-Verlag.

Brillouin, L. (1962) *Science and Information Theory*, 2nd edn. New York: Dover.

Britzke, E. R., and Herzog, C. (2009) *Using Acoustic Surveys to Monitor Population Trends in Bats.* Vicksburg, MS: US Army Engineer Research and Development Center.

Buckland, S. T., Marsden, S. J., and Green, R. (2008) Estimating bird abundance: making methods work, *Bird Conservation International*, **18**, 91–108.

Buckland, S. T., Anderson, D. R., Burnham, K. P., Laake, J. L., Borchers, D. L., and Thomas, L. (2001) *Introduction to Distance Sampling.* Oxford: Oxford University Press.

Bulmer, M. G. (1974) On fitting the Poisson lognormal distribution to species-abundance data, *Biometrics*, **30**, 101–110.

Bunge, J. A. (2013) A survey of software for fitting capture-recapture models, *WIREs Computational Statistics*, **5**, 114–120.

Burton, A. C., Neilson, E. W., Moreira, D., Ladle, A., Steenweg, R., Fisher, J. T., Bayne, E., and Boutin, S. (2015) Wildlife camera trapping: a review and recommendations for linking surveys to ecological processes, *Journal of Applied Ecology*, **52**, 675–685.

Byth, K. (1982) On robust distance-based intensity estimators, *Biometrics*, **38**, 127–135.

Byth, K., and Ripley, B. D. (1980) On sampling spatial patterns by distance methods, *Biometrics*, **36**, 279–284.

Caldwell, Z. R., Zgliczynski, B. J., Williams, G. J., and Sandin, S. A. (2016) Reef fish survey techniques: assessing the potential for standardizing methodologies, *PLoS ONE*, **11**, e0153066.

Camp, R. J., LaPointe, D. A., Hart, P. J., Sedgwick, D. E., and Canale, L. K. (2019) Large-scale tree mortality from Rapid Ohia Death negatively influences avifauna in lower Puna, Hawaii Island, USA, *The Condor*, **121**, duz007.

Canfield, R. H. (1941) Application of the line interception method in sampling range vegetation, *Journal of Forestry*, **39**, 388–394.

Cappo, M., Harvey, E. S., and Shortis, M. R. (2006) Counting and measuring fish with baited video techniques – an overview, *Australian Society for Fish Biology 2006 Workshop Proceedings*, 101–114.

Catana, A. J. (1953) The wandering quarter method of estimating population density, *Ecology*, **44**, 349–360.

Chao, A. (1984) Nonparametric estimation of the number of classes in a population, *Scandinavian Journal of Statistics*, **11**, 265–270.

Chao, A. (1987) Estimating the population size for capture-recapture data with unequal capture probabilities, *Biometrics*, **43**, 783–791.

Chao, A. (1989) Estimating population size for sparse data in capture-recapture experiments, *Biometrics*, **45**, 427–438.

Chao, A. (2001) An overview of closed capture-recapture models, *Journal of Agricultural, Biological and Environmental Statistics*, **6**, 158–175.

Chao, A. (2005) Species richness estimation. In N. Balakrishnan, C. B. Read, and B. Vidakovic (eds), *Encyclopaedia of Statistical Sciences*, 2nd edn. New York: Wiley. vol. 12, pp. 7907–7916.

Chao, A., and Lee, S.-M. (1992) Estimating the number of classes via sample coverage, *Journal of the American Statistician Association*, **87**, 210–217.

Chao, A., and Shen, T.-J. (2010) *Program SPADE (Species Prediction And Diversity Estimation). Program and Users Guide*. Accessed at: http://chao.stat.nthu.edu.tw.

Chao, A., Lee, S.-M., and Jeng, S.-L. (1992) Estimating population size for capture-recapture data when capture probabilities vary by time and individual animal, *Biometrics*, **48**, 201–216.

Chao, A., Wang, Y. T., and Jost, L. (2013) Entropy and the species accumulation curve: a novel entropy estimator via discovery rates of new species, *Methods in Ecology and Evolution*, **4**, 1091–1100.

Chao, A., Chazdon, R. L., Colwell, R. K., and Shen, T.-J. (2006) Abundance-based similarity indices and their estimation when there are unseen species in samples, *Biometrics*, **62**, 361–371.

Chao, A., Gotelli, N. J., Hsieh, T. C., Sander, E. L., Ma, K. H., Colwell, R. K., and Ellison, A. M. (2014) Rarefaction and extrapolation with Hill numbers: a framework for sampling and estimation in species diversity studies, *Ecological Monographs*, **84**, 45–67.

Chao, A., Hsieh, T. C., Chazdon, R. L., Colwell, R. K., and Gotelli, N. J. (2015) Unveiling the species-rank abundance distribution by generalizing the Good-Turing sample coverage theory, *Ecology*, **96**, 1189–1201.

Chapman, D. G. (1951) Some properties of the hypergeometric distribution with applications to zoological censuses, *University of California Publications in Statistics*, **1**, 131–160.

Chapman, D. G. (1954) The estimation of biological populations, *The Annals of Mathematical Statistics*, **25**, 1–15.

Chiarucci, A., Wilson, J. B., Anderson, B. J., and de Dominicis, V. (1999) Cover *versus* biomass as an estimate of species abundance: does it make a difference to the conclusions? *Journal of Vegetation Science*, **10**, 35–42.

Chiu, C.-H., Wang, Y. T., Walther, B. A., and Chao, A. (2014) An improved nonparametric lower bound of species richness via a modified Good-Turing frequency formula, *Biometrics*, **70**, 671–682.

Cintrón, G., and Schaeffer-Novelli, Y. (1984) Methods for studying mangrove structure. In S. C. Snedaker and J. G. Snedaker (eds), *The Mangrove Ecosystem: Research Methods*. Paris: UNESCO. ch. 6.

Clark, P. J., and Evans, F. C. (1954) Distance to nearest neighbor as a measure of spatial relationships in populations, *Ecology*, **35**, 445–453.

Clarke, K. R. (1990) Comparisons of dominance curves, *Journal of Experimental Marine Biology and Ecology*, **138**, 143–157.

Clarke, K. R., and Warwick, R. M. (1999) The taxonomic distinctness measure of biodiversity: weighting of step lengths between hierarchical levels, *Marine Ecology Progress Series*, **184**, 21–29.

Colwell, R. K., and Coddington, J. A. (1994) Estimating terrestrial biodiversity through extrapolation, *Philosophical Transactions of the Royal Society of London, Series B*, **345**, 101–118.

Cook, R. D., and Jacobson, J. O. (1979) A design for estimating visibility bias in aerial surveys, *Biometrics*, **35**, 735–742.

Corlatti, L., Nelli, L., Bertolini, M., Zibordi, F., and Pedrotti, L. (2017) A comparison of four different methods to estimate population size of Alpine marmot (*Marmota marmota*), *Hystrix*, **28**, 1–7.

Cormack, R. M. (1964) Estimates of survival from sighting of marked animals, *Biometrika*, **51**, 429–438.

Cottam, G., and Curtis, J. T. (1956) The use of distance measures in phytosociological sampling, *Ecology*, **37**, 451–460.

Cottam, G., Curtis, J. T., and Hale, B. W. (1953) Some sampling characteristics of a population of randomly dispersed individuals, *Ecology*, **34**, 741–757.

Damgaard, C. (2014) Estimating mean plant cover from different types of cover data: a coherent statistical framework, *Ecosphere*, **5(2)**, 20.

Darroch, J. N., and Ratcliff, D. (1980) A note on capture-recapture estimation, *Biometrics*, **36**, 149–153.

Daubenmire, R. F. (1959) A canopy-cover method of vegetational analysis, *Northwest Science*, **33**, 43–46.

Dawe, E. G., Hoenig, J. M., and Xu, X. (1993) Change-in-ratio and index-removal methods for population assessment and their application to snow crab (*Chionoecetes opilio*), *Canadian Journal of Fisheries and Aquatic Sciences*, **50**, 1467–1476.

Duchesne, R. R., Chopping, M. J., and Tape, K. D. (2016) Capability of the CANAPI algorithm to derive shrub structural parameters from satellite imagery in the Alaskan Arctic, *Polar Record*, **52**, 124–133.

de Vries, P. G. (1973) *A General Theory on Line Intersect Sampling with Application to Logging Residue Inventory*. Wageningen, Netherlands: Mededelingen Landbouwhogeschool.

DeLury, D. B. (1947) On the estimation of biological populations, *Biometrics*, **3**, 145–167.

Denney, C., Fields, R., Gleason, M., and Starr, R. (2017) Development of new methods for quantifying fish density using underwater stereo-video tools, *Journal of Visualized Experiments*, **129**, 56635.

Diggle, P. J. (1975) Robust density estimation using distance methods, *Biometrika*, **62**, 39–48.

Diggle, P. J. (1983) *Statistical Analysis of Spatial Point Patterns*. London: Academic Press.

Dobrowski, S. Z., and Murphy, S. K. (2006) A practical look at the variable area transect, *Ecology*, **87**, 1856–1860.

Dollar, S. J. (1982) Wave stress and coral community structure in Hawaii, *Coral Reefs*, **1**, 71–81.

Domin, K. (1928) The relations of the Tatra mountain vegetation to the edaphic factors of the habitat: a synecological study, *Acta Botanica Bohemica*, **6/7**, 133–164.

Dorazio, R. M., Jelks, H. L., and Jordan, F. (2005) Improving removal-based estimates of abundance by sampling a population of spatially distinct subpopulations, *Biometrics*, **61**, 1093–1101.

Ducey, M. J., Gove, J. H., and Valentine, H. T. (2004) A walkthrough solution to the boundary overlap problem, *Forest Science*, **50**, 427–435.

Ducey, M. J., Williams, M. S., Gove, J. H., Roberge, S., and Kenning, R. S. (2013) Distance-limited perpendicular distance sampling for coarse woody debris: theory and field results, *Forestry*, **86**, 119–128.

Duchamp, J. E., Yates, M., Muzika, R.-M., and Swihart, R. K. (2006) Estimating probabilities of detection for bat echolocation calls: an application of the double-observer method, *Wildlife Society Bulletin*, **34**, 408–412.

Efford, M. G. (2004) Density estimation in live-trapping studies, *Oikos*, **106**, 598–610.

Efford, M. G. (2011) Estimation of population density by spatially explicit capture-recapture with area searches, *Ecology*, **92**, 2202–2207.

Efford, M. G. (2019) Non-circular home ranges and the estimation of population density, *Ecology*, **100**, e02580. doi:10.1002/ecy.2580.

Elzinga, C. L., Salzer, D. W., and Willoughby, J. W. (1998) *Measuring and Monitoring Plant Populations*. Denver, CO: Bureau of Land Management.

Engeman, R. M., Nielsen, R. M., and Sugihara, R. T. (2005) Evaluation of optimized variable area transect sampling using totally enumerated field data sets, *Environmetrics*, **16**, 767–772.

Engeman, R. M., Sugihara, R. T., Pank, L. F., and Dusenberry, W. E. (1994) A comparison of plotless density estimators using Monte Carlo simulation, *Ecology*, **75**, 1769–1779.

Enquist, B. J., Feng, X., Boyle, B., Maltner, B., Newman, E. A., Jørgensen, P. M., Roehrdanz, P. R., Thiers, B. M., Burger, J. R., Corlett, R. T., Couvreur, L. P., Dauby, G., Donoghue, J. C., Foden, W., Lovett, J. C., Marquet, P. A., Merow, C., Midgley, G., Morueta-Holme, N., Neves, D. M., Oliveira-Filho, A. T., Kraft, N. J. B., Park, D. S., Peet, R. K., Pillet, M., Serra-Diaz, J. M., Sandel, B., Schildhauer, M., Šimová, I., Violle, C., Wieringa, J. J., Wiser, S. K., Hannah, L., Svenning, J.-C., and McGill, B. J. (2019) The commonness of rarity: global and future distribution of rarity across land plants, *Science Advances*, **5**, eaaz0414.

Espeland, E. K., and Emam, T. M. (2011) The value of structuring rarity: the seven types and links to reproductive ecology, *Biodiversity and Conservation*, **20**, 963–985.

Evans, R. A., and Love, R. M. (1957) The step-point method of sampling: a practical tool in range research, *Journal of Range Management*, **10**, 208–212.

Farnsworth, G. I., Pollock, K. H., Nichols, J. D., Simons, T. R., Hines, J. E., and Sauer, J. R. (2002) A removal model for estimating detection probabilities from point-count surveys, *The Auk*, **119**, 414–425.

Fattorini, S., Cardoso, P., Rigal, F., and Borges, P. A. V. (2012) Use of arthropod rarity for area prioritisation: insights from the Azorean islands, *PLoS ONE*, **7**, e33995.

Fehmi, J. S. (2010) Confusion among three common plant cover definitions may result in data unsuited for comparison, *Journal of Vegetation Science*, **21**, 273–279.

Fernandes, P. G., Ferlotto, F., Holliday, D. V., Nakken, O., and Simmonds, J. (2002) Acoustic applications in fisheries science: the ICES contribution, *ICES Marine Symposia*, **215**, 483–492.

Fisher, R. A., Corbet, A. S., and Williams, C. B. (1943) The relation between the number of species and the number of individuals in a random sample of an animal population, *Journal of Animal Ecology*, **12**, 42–58.

Fiske, I. J., and Chandler, R. B. (2011) unmarked: an R package for fitting hierarchical models of wildlife occurrence and abundance, *Journal of Statistical Software*, **43**, 1–23.

Fletcher, R. J., Jr, and Hutto, R. L. (2006) Estimating detection probabilities of river birds using double surveys, *The Auk*, **123**, 695–707.

Forcey, G. M., and Anderson, J. T. (2002) Variation in bird detection probabilities and abundances among different point count durations and plot sizes, *Proceedings of the Annual Conference of the Southeastern Association of Fish and Wildlife Agencies*, **56**, 331–342.

Forcey, G. M., Anderson, J. T., Ammer, F. K., and Whitmore, R. C. (2006) Comparison of two double-observer point-count approaches for estimating breeding bird abundance, *Journal of Wildlife Management*, **70**, 1674–1681.

Gibbons, D. W., and Gregory, R. D. (2006) Birds. In W. J. Sutherland (ed.), *Ecological Census Techniques*. Cambridge: Cambridge University Press. ch. 9.

Good, I. J. (1953) The population frequencies of species and the estimation of population parameters, *Biometrika*, **40**, 237–264.

Goodall, D. W. (1952) Some considerations in the use of point quadrats for the analysis of vegetation, *Australian Journal of Scientific Research, Series B*, **5**, 1–41.

Gove, J. H., and van Deusen, P. C. (2011) On fixed-area plot sampling for downed coarse woody debris, *Forestry*, **84**, 109–117.

Gove, J. H., Ducey, M. J., Valentine, H. T., and Williams, M. S. (2013) A comprehensive comparison of perpendicular distance methods for sampling downed coarse wood debris, *Forestry*, **86**, 129–143.

Gray, J. S., Bjørgesæter, M. K., and Ugland, K. I. (2006) On plotting species abundance distributions, *Journal of Animal Ecology*, **75**, 752–756.

Gregoire, T. G. (1982) The unbiasedness of the mirage correction procedure for boundary overlap, *Forest Science*, **28**, 504–508.

Gregoire, T. G., and Valentine, H. T. (2003) Line intersect sampling: ell-shaped transects and multiple intersections, *Environmental and Ecological Statistics*, **10**, 263–279.

Grimm, A., Gruber, B., and Henle, K. (2014) Reliability of different mark-recapture methods for population size estimation tested against reference population sizes constructed from field data, *PLoS ONE*, **9**, e98840.

Grosenbaugh, L. R. (1964) Some suggestions for better sample-tree-measurement. In *Proceedings of the Society of American Foresters Annual Meeting*, 20–23 October 1963, Boston. pp. 36–42.

Guillera-Arrolta, G., Kéry, M., and Lahoz-Monfort, J. J. (2019) Inferring species richness using multispecies occupancy modeling: estimation performance and interpretation, *Ecology and Evolution*, **9**, 780–792.

Halford, A. R., and Thompson, A. A. (1994) *Visual Census Surveys of Reef Fish*. Townsville: Australian Institute of Marine Science.

Hall, P. G., Melville, G. J., and Welsh, A. H. (2001) Bias correction and bootstrap methods for a spatial sampling scheme, *Bernoulli*, 7, 829–846.

Harley, S. J., Myers, R. A., and Dunn, A. (2001) Is catch-per-unit-effort proportional to abundance? *Canadian Journal of Fisheries and Aquatic Sciences*, 58, 1760–1772.

Harrison, S. P., Ross, S. J., and Lawton, J. H. (1992) Beta diversity on geographic gradients in Britain, *Journal of Animal Ecology*, 61, 151–158.

Hartley, L. J. (2012) Five-minute bird counts in New Zealand, *New Zealand Journal of Ecology*, 36, 268–278.

Hartley, S., and Kunin, W. E. (2003) Scale dependency of rarity, extinction risk, and conservation priority, *Conservation Biology*, 17, 1559–1570.

Harvey, E. S., Cappo, M., Butler, J. J., Hall, N., and Kendrick, G. A. (2007) Bait attraction affects the performance of remote underwater video stations in assessment of demersal fish community structure, *Marine Ecology Progress Series*, 350, 245–254.

He, F., and Gaston, K. J. (2000) Estimating species abundance from occurrence, *The American Naturalist*, 156, 553–559.

Heard, G. W., Scroggie, M. P., Clemann, N., and Ramsey, D. S. L. (2014) Wetland characteristics influence disease risk for a threatened amphibian, *Ecological Applications*, 24, 250–262.

Heip, C. (1974) A new index measuring evenness, *Journal of the Marine Biological Association of the United Kingdom*, 54, 555–557.

Hijbeek, R., Koedam, N., Khan, M. N. I., Kairo, J. G., Schoukens, J., and Dahdouh-Guebas, F. (2013) An evaluation of plotless sampling using vegetation simulations and field data from a mangrove forest, *PLoS ONE*, 8, e67201.

Hill, M. O. (1973) Diversity and evenness: a unifying notation and its consequences, *Ecology*, 54, 427–432.

Horvitz, D. G., and Thompson, D. J. (1952) A generalization of sampling without replacement from a finite universe, *Journal of the American Statistical Association*, 47, 663–685.

Huggins, R. M. (1989) On the statistical analysis of capture experiments, *Biometrika*, 76, 133–140.

Jennings, S. B., Brown, N. D., and Sheil, D. (1999) Assessing forest canopies and understorey illumination: canopy closure, canopy cover and other measures, *Forestry*, 72, 59–74.

Johannesson, K. A., and Mitson, R. B. (1983) Fisheries acoustics: a practical manual for aquatic biomass estimation, *FAO Fisheries Technical Paper*, 240, 1–249.

Jolly, G. M. (1963) Estimates of population parameters from multiple recapture data with both death and dilution – deterministic model, *Biometrika*, 50, 113–128.

Jolly, G. M. (1965) Explicit estimates from capture-recapture data with both death and immigration-stochastic model, *Biometrika*, 52, 225–247.

Jost, L. (1993) A simple distance estimator for plant density in nonuniform stands: mathematical appendix. Accessed at: http://www.loujost.com/Statistics%20and%20Physics/PCQ/PCQJournalArticle.htm.

Kaiser, L. (1983) Unbiased estimation in line-intercept sampling, *Biometrics*, 39, 965–976.

Keating, K. A., Schwartz, C. C., Haroldson, M. A., and Moody, D. (2002) Estimating numbers of females with cubs-of-the-year in the Yellowstone grizzly bear population, *Ursus*, 13, 161–174.

Keeley, J. E., and Fotheringham, C. J. (2005) Plot shape effects on plant species diversity measurements, *Journal of Vegetation Science*, 16, 249–256.

Kelker, G. H. (1940) Estimating deer populations by a differential hunting loss in the sexes, *Proceedings of the Utah Academy of Science*, 17, 65–69.

Kemp, C. D., and Kemp, A. W. (1956) The analysis of point quadrat data, *Australian Journal of Botany*, 4, 167–174.

Kendall, W. L., and Bjorkland, R. (2001) Using open robust design models to estimate temporary emigration from capture-recapture data, *Biometrics*, 57, 1113–1122.

Kendall, W. L., Nichols, J. D., and Hines, J. E. (1997) Estimating temporary emigration using capture-recapture data with Pollocks robust design, *Ecology*, **78**, 563–578.

Kendall, W. L., Pollock, K. H., and Brownie, C. (1995) A likelihood-based approach to capture-recapture estimation of demographic parameters under the robust design, *Biometrics*, **51**, 293–308.

Kent, M., and Coker, P. (1994) *Vegetation Description and Analysis*. Chichester: Wiley.

Kershaw, J. A., Jr, Ducey, M. J., Beers, T. W., and Husch, B. (2016) *Forest Mensuration*, 5th edn. New York: Wiley-Blackwell.

Khan, M. N. I., Hijbeek, R., Berger, U., Koedam, N., Grueters, U., Islam, S. M. Z., Hasan, M. A., and Dahdouh-Guebas, F. (2016) An evaluation of the plant density estimator the Point-Centred Quarter Method (PCQM) using Monte Carlo simulation, *PLoS ONE*, **11(6)**, e0157985.

Kiani, B., Fallah, A., Tabari, M., Hosseini, S. M., and In Parizi, M.-H. (2013) A comparison of distance sampling methods in Saxaul (*Halloxylon Ammodendron* c. a. Mey Bunge) shrublands, *Polish Journal of Ecology*, **61**, 207–219.

Koleff, P., Gaston, K. J., and Lennon, J. J. (2003) Measuring beta diversity for presence-absence data, *Journal of Animal Ecology*, **72**, 367–382.

Krajina, V. J. (1933) Die pflanzengesellschaften des Mlynica-Tales in den Visoke Tatry (Hohe Tatra), *Beihefte zum Botanischen Centralblatt*, **50**, 774–957; **51**, 1–224.

Kuehl, R. O., McClaran, M. P., and van Zee, J. (2001) Detecting fragmentation of cover in desert grasslands using line intercept, *Journal of Range Management*, **54**, 61–66.

Kvålseth, T. O. (2015) Evenness indices once again: critical analysis of properties, *SpringerPlus*, **4**, 232.

Leinster, T., and Cobbold, C. A. (2012) Measuring diversity: the importance of species similarity, *Ecology*, **93**, 477–489.

Lennon, J. J., Koleff, P., Greenwood, J. J. D., and Gaston, K. J. (2001) The geographical structure of British bird distributions: diversity, spatial turnover and scale, *Journal of Animal Ecology*, **70**, 966–979.

Leslie, P. H., and Davis, D. H. S. (1939) An attempt to determine the absolute number of rats on a given area, *Journal of Animal Ecology*, **8**, 94–113.

Leujak, W., and Ormond, R. F. G. (2007) Comparative accuracy and efficiency of six coral community survey methods, *Journal of Experimental Marine Biology and Ecology*, **351**, 168–187.

Levy, E. B., and Madden, E. A. (1933) The point method of pasture analysis, *New Zealand Journal of Agriculture*, **46**, 267–279.

Lincoln, F. C. (1930) *Calculating Waterfowl Abundance on the Basis of Banding Returns*. Circular 118. Washington, DC: US Department of Agriculture.

Loeb, S. C., Rodhouse, T. J., Ellison, L. E., Lausen, C. L., Reichard, J. D., Irvine, K. M., Ingersoll, T. E., Coleman, J. T. H., Thogmartin, W. E., Sauer, J. R., Francis, C. M., Bayless, M. L., Stanley, T. T., and Johnson, D. H. (2015) *A Plan for the North American Bat Monitoring Program (NABat)*. Asheville, NC: USDA.

Longino, J. T., Coddington, J., and Colwell, R. K. (2002) The ant fauna of a tropical rain forest: estimating species richness three different ways, *Ecology*, **83**, 689–702.

Lucas H. A., and Seber, G. A. F. (1977) Estimating coverage and particle density using the line intercept method, *Biometrika*, **64**, 618–622.

Mac Nally, R., Duncan, R. P., Thomson, J. R., and Yen, J. D. L. (2017) Model selection using information criteria, but is the 'best' model any good? *Journal of Applied Ecology*, **55**, 1441–1444.

MacArthur, R. H. (1965) Patterns of species diversity, *Biological Reviews*, **40**, 510–533.

Magurran, A. E., Queiroz, H. L., and Hercos, A. P. (2013) Relationship between evenness and body size in species rich assemblages, *Biology Letters*, **9**. Accessed at: https://doi.org/10.1098/rsbl.2013.0856.

Mallet, D., and Pelletier, D. (2014) Underwater video techniques for observing coastal marine biodiversity: a review of sixty years of publications (1952–2012), *Fisheries Research*, **154**, 44–62.

Manly, B. F. J. (1984) Obtaining confidence limits on parameters of the Jolly-Seber model for capture-recapture data, *Biometrics*, **40**, 749–758.

Manly, B. F. J., and McDonald, L. L. (1996) Sampling wildlife populations, *Chance*, **9**, 9–20.

Margalef, R. (1958) Information theory in ecology, *International Journal of General Systems*, **3**, 36–71.

Marini, S., Fanelli, E., Sbragaglia, V., Azzurro, E., del Rio Fernandez, J., and Aguzzi, J. (2018) Tracking fish abundance by underwater image recognition, *Scientific Reports*, **8**, 137–148.

Mark, A. F., and Esler, A. E. (1970) An assessment of the point-centred quarter method of plotless sampling in some New Zealand forests, *Proceedings of the New Zealand Ecological Society*, **17**, 106–110.

Marshall, D. D., Iles, K., and Bell, J. F. (2004) Using a large-angle gauge to select trees for measurement in variable plot sampling, *Canadian Journal of Forestry Research*, **34**, 840–845.

Masuyama, M. (1954) On the error in crop cutting experiment due to the bias on the border of the grid, *Sankhya*, **14**, 181–186.

Matheron, G. (1989) *Estimating and Choosing*. Berlin: Springer.

Mathews, F., Kubasiewicz, L. M., Gurnell, J., Harrower, C. A., McDonald, R. A., and Shore, R. F. (2018) *A Review of the Population and Conservation Status of British Mammals: Technical Summary*. A report by the Mammal Society under contract to Natural England, Natural Resources Wales and Scottish Natural Heritage. Peterborough: Natural England.

Matsuoka, S. M., Mahon, C. L., Handel, C. M., Sólymos, P., Bayne, E. M., Fontaine, P. C., and Ralph, C. J. (2014) Reviving common standards in point-count surveys for broad inference across studies, *The Condor*, **116**, 599–608.

Matthews, T. J., Borregaard, M. K., Ugland, K. I., Borges, P. A. V., Rigal, F., Cardoso, P., and Whittaker, R. J. (2014) The gambin model provides a superior fit to species abundance distributions with a single free parameter: evidence, implementation and interpretation, *Ecography*, **37**, 1002–1011.

Matthews, T. J., Triantis, K. A., Rigal, F., Borregaard, M. K., Guilhaumon, F., and Whittaker, R. J. (2016) Island species-area relationships and species accumulation curves are not equivalent: an analysis of habitat island datasets, *Global Ecology and Biogeography*, **25**, 607–618.

Matthews, T. J., Borges, P. A. V., Azevedo, E. B., and Whittaker, R. J. (2017) A biogeographical perspective on species abundance distributions: recent advances and opportunities for future research, *Journal of Biogeography*, **44**, 1705–1710.

Matthews, T. J., Borregaard, M. K., Gillespie, C. S., Rigal, F., Ugland, K. I., Krüger, R. F., Marques, R., Sadler, J. P., Borges, P. A. V., Kubota, Y., and Whittaker, R. J. (2019) Extension of the gambin model to multimodal species abundance distributions, *Methods in Ecology and Evolution*, **10**, 432–437.

Maunder, M. N., Sibert, J. R., Fonteneau, A., Hampton, J., Kleiber, P., and Harley, S. J. (2006) Interpreting catch per unit effort data to assess the status of individual stocks and communities, *ICES Journal of Marine Science*, **63**, 1373–1385.

McClintock, B. T., White, G. C., and Pryde, M. A. (2009) Improved methods for estimating abundance and related demographic parameters from mark-resight data, *Biometrics*, **75**, 1–11.

McClintock, B. T., White, G. C., Antolin, M. F., and Tripp, D. W. (2009) Estimating abundance using mark-resight when sampling is with replacement or the number of marked individuals is unknown, *Biometrics*, **65**, 237–246.

McDonald, L. L. (1980) Line-intercept sampling for attributes other than coverage and density, *Journal of Wildlife Management*, **44**, 530–533.

McGill, B. J., Etienne, R. S., Gray, J. S., Alonso, D., Anderson, M. J., Benecha, H. K., Dornelas, M., Enquist, B. J., Green, J. L., He, F., Hurlbert, A. H., Magurran, A. E., Marquet, P. A.,

Maurer, B. A., Ostling, A., Soykan, C. U., Ugland, K. I., and White, E. P. (2007) Species abundance distributions: moving beyond single prediction theories to integration within an ecological framework, *Ecology Letters*, **10**, 995–1015.

Menhinick, E. F. (1964) A comparison of some species-individuals diversity indices applied to samples of field insects, *Ecology*, **45**, 859–861.

Mitchell, K. (2015) *Quantitative Analysis by the Point-Centered Quarter Method*. Accessed at: https://arxiv.org/abs/1010.3303v2.

Moore, P. G. (1954) Spacing in plant populations, *Ecology*, **35**, 222–227.

Moran, P. A. P. (1951) A mathematical theory of animal trapping, *Biometrika*, **38**, 307–311.

Morisita, M. (1954) Estimation of population density by spacing method, *Memoirs of the Faculty of Science, Kyushu University, Series E*, **1**, 187–197.

Morisita, M. (1957) A new method for the estimation of density by the spacing method applicable to non-randomly distributed populations (translated by USDA, Forest Service, 1960), *Physiology and Ecology* (in Japanese), **7**, 134–144.

Morisita, M. (1959) Measuring of the dispersion and analysis of distribution patterns, *Memoires of the Faculty of Science, Kyushu University, Series E. Biology*, **2**, 215–235.

Morrison, D. A., Le Brocque, A. F., and Clarke, P. J. (1995) An assessment of some improved techniques for estimating the abundance (frequency) of sedentary organisms, *Vegetatio*, **120**, 131–145.

Nemec, A. F. L., and Davis, G. (2002) Efficiency of six line intersect sampling designs for estimating volume and density of coarse woody debris. *Research Section, Vancouver Forest Region, B.C. Ministry of Forests, Nanaimo, B.C., Technical Report, TR-021/2002*.

Nichols, J. D., Hines, J. E., Sauer, J. R., Fallon, F. W., Fallon, J. E., and Heglund, P. J. (2000) A double-observer approach for estimating detection probability and abundance from point counts, *The Auk*, **117**, 393–408.

Niedballa, J., Sollmann, R., Courtiol, A., and Wilting, A. (2016) camtrapR: an R package for efficient camera trap data management, *Methods in Ecology and Evolution*, **7**, 1457–1462.

Norvell, R. E., Howe, F. P., and Parrish, J. R. (2003) A seven-year comparison of relative-abundance and distance-sampling methods, *The Auk*, **120**, 1013–1028.

Otis, D. L., Burnham, K. P., White, G. C., and Anderson, D. R. (1978) Statistical inference from capture data on closed animal populations, *Wildlife Monographs*, **62**, 11–35.

Outhred, R. K. (1984) Semi-quantitative sampling in vegetation survey. In K. Myers, C. R. Margules and I. Musto (eds), *Survey Methods for Nature Conservation*. Canberra: CSIRO. pp. 87–100.

Palmer, M. W., and White, P. S. (1994) Scale dependence and the species-area relationship, *The American Naturalist*, **144**, 717–740.

Parker, K. R. (1979) Density estimation by variable area transect, *Journal of Wildlife Management*, **43**, 484–492.

Peet, R. K., Wentworth, T. R., and White, P. S. (1998) A flexible, multipurpose method for recording composition and structure, *Castanea*, **63**, 262–274.

Petersen, C. G. J. (1896) The yearly immigration of young plaice into the Limfjord from the German Sea, *Report of the Danish Biological Station*, **6**, 5–84.

Pielou, E. C. (1966) The measurement of diversity in different types of biological collections, *Journal of Theoretical Biology*, **13**, 131–144.

Pielou, E. C. (1979) *Mathematical Ecology*. New York: Wiley.

Pollard, J. H. (1971) On distance estimators of density in randomly distributed forests, *Biometrics*, **27**, 991–1002.

Pollock, K. H. (1982) A capture-recapture design robust to unequal probability of capture, *Journal of Wildlife Management*, **46**, 757–760.

Pollock, K. H., Hines, J. E., and Nichols, J. D. (1984) The use of auxiliary variables in capture-recapture and removal experiments, *Biometrics*, **40**, 329–340.

Preston, F. W. (1948) The commonness, and rarity, of species, *Ecology*, **29**, 254–283.

Prodan, M. (1968) Punktstichprobe für die Forsteinrichtung, *Der Forst- und Holzwirt*, **23**, 225–226.

Rabinowitz, D. (1981) Seven forms of rarity. In H. Synge (ed.), *The Biological Aspects of Rare Plant Conservation*. Chichester: Wiley. pp. 205–217.

Rademaker, M., Meijaard, E., Semiadi, G., Blokland, S., Neilson, E. W., and Rode-Margono, E. J. (2016) First ecological study of the Bawean Warty Pig (*Sus blouchi*), one of the rarest pigs on Earth, *PLoS ONE*, **11**, e0151732.

Ralph, C. J., Droege, S., and Sauer, J. R. (1995) Managing and monitoring birds using point counts: standards and applications. In C. J. Ralph, J. R. Sauer, and S. Droege (eds), *Monitoring Bird Populations by Point Counts*. Albany, CA: US Forest Service. pp. 161–175.

Rao, C. R. (1982) Diversity and dissimilarity coefficients: a unified approach, *Theoretical Population Biology*, **21**, 24–43.

Rao, C. R. (2010) Quadratic entropy and analysis of diversity, *Sankhyā*, **72-A**, 70–80.

Rayburn, A. P., Schiffers, K., and Schupp, E. W. (2011) Use of precise spatial data for describing spatial patterns and plant interactions in a diverse Great Basin shrub community, *Plant Ecology*, **212**, 585–594.

Riddle, J. D., Pollock, K. H., and Simons, T. R. (2010) An unreconciled double-observer method for estimating detection probability and abundance, *The Auk*, **127**, 841–849.

Rivest, L.-P., and Baillargeon, S. (2007) Applications and extensions of Chao's moment estimator for the size of a closed population, *Biometrics*, **63**, 999–1006.

Roberts, T. E., Bridge, T. C., Caley, M. J., Baird, A. H. (2016) The point count transect method for estimates of biodiversity on coral reefs: improving the sampling of rare species. *PLoS ONE*, **11**, e0152335.

Rohlf, F. J., and Archie, J. W. (1978) Least-squares mapping using interpoint distances, *Ecology*, **59**, 126–132.

Rothery, P. (1974) The number of pins in a point quadrat frame, *Journal of Applied Ecology*, **11**, 745–754.

Rowcliffe, J. M., Field, J., Turvey, S. T., and Carbone, C. (2008) Estimating animal density using camera traps without the need for individual recognition, *Journal of Animal Ecology*, **45**, 1228–1236.

Rowcliffe, J. M., Carbone, C., Jansen, P. A., Kays, R., and Kranstauber, B. (2011) Quantifying the sensitivity of camera traps: an adapted distance sampling approach, *Methods in Ecology and Evolution*, **2**, 464–476.

Royle, J. A. (2004) Generalized estimators of avian abundance from count survey data, *Animal Biodiversity and Conservation*, **27**, 375–386.

Royle, J. A., and Nichols, J. D. (2003) Estimating abundance from repeated presence-absence data or point counts, *Ecology*, **84**, 777–790.

Royle, J. A., and Young, K. V. (2008) A hierarchical model for spatial capture-recapture data, *Ecology*, **89**, 2281–2289.

Royle, J. A., Fuller, A. K., and Sutherland, C. (2016) Spatial capture-recapture models allowing Markovian transience or dispersal, *Population Ecology*, **58**, 53–62.

Russell, M. B., Fraver, S., Aakala, T., Gove, J. H., Woodall, C. W., D'Amato, A. W., and Ducey, M. J. (2015), Quantifying carbon stores and decomposition in dead wood: A review, *Forest Ecology and Management*, **350**, 107–128.

Sadinle, M. (2009) Transformed logit confidence intervals for small populations in single capture-recapture estimation, *Communications in Statistics – Simulation and Computation*, **38**, 1909–1924.

Schmid-Haas, P. (1969) Stichproben am waldrand, *Mitteilungen der Schweizerische Anstalt für das Forstliche Versuchswesen*, **45**, 234–303.

Schmidt, J. H., and Rattenbury, K. L. (2017) An open-population distance sampling framework for assessing population dynamics in group-dwelling species, *Methods in Ecology and Evolution*, **9**, 936–945.

Schnabel, Z. E. (1938) The estimation of total fish population of a lake, *The American Mathematical Monthly*, **45**, 348–352.

Seber, G. A. F. (1965) A note on the multiple recapture census, *Biometrika*, **52**, 249–259.

Seber, G. A. F. (1982) *The Estimation of Animal Abundance and Related Parameters*, 2nd edn. London: Griffin.

Shannon, C. E. (1948) A mathematical theory of communication, *The Bell System Technical Journal*, **27**, 379–423.

Sheldon, A. L. (1969) Equitability indices: dependence on the species count, *Ecology*, **50**, 466–467.

Sheil, D. (2002) Biodiversity research in Malinau. In *CIFOR, ITTO Project PD 12/97 Rev. I (F): Forest, science and sustainability: the Bulungan model forest. Technical Report Phase 1, 1997–2001.* Bogor, Indonesia: CIFOR and ITTO. pp. 57–107.

Sheil, D., Ducey, M. J., Sidiyasa, K., and Samsoedin, I. (2003) A new type of sample unit for the efficient assessment of diverse tree communities in complex forest landscapes, *Journal of Tropical Forest Science*, **15**, 117–135.

Silva, L. B., Alves, M., Elias, R. B., and Silva, L. (2017) Comparison of T-square, point-centered quarter, and N-tree sampling methods in *Pittosporum undulatum* invaded woodlands, *International Journal of Forestry Research*, 2818132. Accessed at: https://doi.org/10.1155/2017/2818132.

Simpson, E. H. (1949) Measurement of diversity, *Nature*, **163**, 688.

Smith, W. P., Twedt, D. J., Hamel, P. B., Ford, R. P., Weidenfeld, D. A., and Cooper, R. J. (1998) Increasing point-count duration increases standard error, *Journal of Field Ornithology*, **69**, 450–456.

Stevens, D. L., Jr, and Olsen, A. R. (2004) Spatially balanced sampling of natural resources, *Journal of the American Statistician Association*, **99**, 262–278.

Stohlgren, T. J., Falkner, M. B., and Schell, L. D. (1995) A Modified-Whittaker nested vegetation sampling method, *Vegetatio*, **117**, 113–121.

Strauss, D., and Neal, D. L. (1963) Biases in the step-point method on bunchgrass ranges, *Journal of Range Management*, **36**, 623–626.

Tidmarsh, C. E. M., and Havenga, C. M. (1955) *The Wheel-Point Method of Survey and Measurement of Semi-Open Grasslands and Karoo Vegetation in South Africa*. Pretoria: Botanical Survey of South Africa, Memoir 29.

Ugland, K. I., Lambshead, P. J. D., McGill, B. J., Gray, J. S., O'Dea, N., Ladle, R. J., and Whittaker, R. J. (2007) Modelling dimensionality in species abundance distributions: description and evaluation of the gambin model, *Evolutionary Ecology Research*, **9**, 313–324.

Upton, G. J. G. (2016) *Categorical Data Analysis by Example*. New York: Wiley.

Upton, G. J. G., and Fingleton, B. (1985) *Spatial Data Analysis by Example. Vol 1: Point Pattern and Quantitative Data*. Chichester: Wiley.

Valentine, H. T., Ducey, M. J., Gove, J. H., Lanz, A., and Affleck, D. L. R. (2006) Corrections for cluster-plot slop, *Forest Science*, **52**, 55–66.

Vicharnakorn, P., Shrestha, R. P., Nagai, M., Salam, A. P., and Kiratiprayoon, S. (2014) Carbon stock assessment using remote sensing and forest inventory data in Savannakhet, Lao PDR, *Remote Sensing*, **6**, 5452–5479.

Warde, W., and Petranka, J. W. (1981) A correction factor table for missing point-center quarter data, *Ecology*, **62**, 491–494.

Warwick, R. M., and Clarke, K. R. (1995) New 'biodiversity' measures reveal a decrease in taxonomic distinctness with increasing stress, *Marine Ecology Progress Series*, **129**, 301–305.

West, P. W. (2011) Potential for wider application of 3P sampling in forest inventory, *Canadian Journal of Forestry Research*, **41**, 1500–1508.

White, G. C., and Burnham, K. P. (1999) Program MARK: survival estimation from populations of marked animals, *Bird Study*, **46**, sup1, S120–S139.

White, N. A., Engeman, R. M., Sugihara, R. T., and Krupa, H. W. (2008) A comparison of plotless density estimators using Monte Carlo simulation on totally enumerated field data sets, *BMC Ecology*, **8**, 6.

Whittaker, R. H. (1960) Vegetation of the Siskiyou mountains, Oregon and California, *Ecological Monographs*, **30**, 279–338.

Whittaker, R. H. (1965) Dominance and diversity in land plant communities, *Science*, **147**, 250–260.

Whittaker, R. H. (1972) Evolution and measurement of species diversity, *Taxon*, **21**, 213–251.

Whittaker, R. H. (1977) Evolution of species diversity on land communities, *Journal of Evolutionary Biology*, **10**, 1–67.

Whittington, J., and Sawaya, M. A. (2015) A comparison of grizzly bear demographic parameters estimated from non-spatial and spatial open population capture-recapture models, *PLoS ONE*, **10**, e0134446.

Widmer, L., Heule, E., Colombo, M., Ruegg, A., Indermaur, A., Ronco, F., and Salzburger, W. (2019) Point-Combination Transect (PCT): incorporation of small underwater cameras to study fish communities, *Methods in Ecology and Evolution*, **10**, 891–901.

Willers, J. L., Yatham, S. R., Williams, M. R., and Akins, D. C. (1992) Utilization of the line-intercept method to estimate the coverage, density, and average length of row skips in cotton and other row crops, *Annual Conference on Applied Statistics in Agriculture, Proceedings*, 48–59.

Williams, F. M. (1977) Model-free evenness: an alternative to diversity measures. Paper S23-6 presented 8/8/1977 to Second International Ecological Congress, Satellite program in Statistical Ecology, Satellite A: NATO-Advanced Study Institute and ISEP Research Workshop, Jerusalem.

Williams, M. S., and Gove, J. H. (2003) Perpendicular distance sampling, another method of sampling coarse woody debris, *Canadian Journal of Forestry Research*, **33**, 1564–1579.

Williams, M. S., Ducey, M. J., and Gove, J. H. (2005) Assessing surface area of coarse woody debris with line intersect and perpendicular distance sampling, *Canadian Journal of Forestry Research*, **35**, 949–960.

Williamson, M., and Gaston, K. J. (2005) The lognormal distribution is not an appropriate null hypothesis for the species-abundance distribution, *Journal of Animal Ecology*, **74**, 409–422.

Willis, T. J. (2001) Visual census methods underestimate density and diversity of cryptic reef fishes, *Journal of Fish Biology*, **59**, 1408–1411.

Wilson, D. J., Mulvey, R. L., Clarke, D. A., and Reardon, J. T. (2017) Assessing and comparing population densities and indices of skinks under three predator management regimes, *New Zealand Journal of Ecology*, **41**, 84–97.

Wilson, J. W. (1960) Inclined point quadrats, *New Phytologist*, **59**, 1–7.

Wilson, J. W. (1963) Errors resulting from thickness of point quadrats, *Australian Journal of Botany*, **11**, 178–188.

Wilson, M. V., and Shmida, A. (1984) Measuring beta diversity with presence-absence data, *Journal of Ecology*, **72**, 1055–1064.

Wilson, R. M., and Collins, M. F. (1992) Capture-recapture estimation with samples of size one using frequency data, *Biometrika*, **79**, 543–553.

Yang, T.-R., Hsu, Y.-H., Kershaw, J. A., Jr, McGarrigle, E., and Kilham, D. (2017) Big BAF sampling in mixed species forest structures of northeastern North America: influence of count and measure BAF under cost constraints, *Forestry*, **90**, 649–660.

Yarranton, G. A. (1966) A plotless method of sampling vegetation, *Journal of Ecology*, **54**, 229–237.

Yee, T. W., Stoklosa, J., and Huggins, R. M. (2015) The VGAM package for capture-recapture data using the conditional likelihood, *Journal of Statistical Software*, **65**, 1–33.

Yen, J. D. L., Thomson, J. R., and Mac Nally, R. (2012) Is there an ecological basis for species abundance distributions?, *Oecologia*, **171**, 517–525.

Yen, J. D. L., Thomson, J. R., Paganin, D. M., Keith, J. M., and Mac Nally, R. (2014) Function regression in ecology and evolution: FREE, *Methods in Ecology and Evolution*, **6**, 17–26.

Yim, J.-S., Shin, M.-Y., Son, Y., and Kleinn, C. (2015) Cluster plot optimization for a large area forest resource inventory in Korea, *Forest Science and Technology*, **11**, 139–146.

Yin, D., and He, F. (2014) A simple method for estimating species abundance from occurrence maps, *Methods in Ecology and Evolution*, **5**, 336–343.

Yu, J., and Dobson, F. S. (2001) Seven forms of rarity in mammals, *Journal of Biogeography*, **27**, 131–139.

Yuan, Y., Buckland, S. T., Harrison, P. J., Foss, S., and Johnston, A. (2016) Using species proportions to quantify turnover in diversity, *Journal of Agricultural, Biological and Environmental Statistics*, **21**, 363–381.

Zarco-Perello, S., and Enríquez, S. (2019) Remote underwater video reveals higher fish diversity and abundance in seagrass meadows, and habitat differences in trophic interactions, *Scientific Reports*, **9**, 6596.

Zvuloni, A., Artzy-Randrup, Y., Stone, L., van Woesik, R., and Loya, Y. (2008) Ecological size-frequency distributions: how to prevent biases in spatial sampling, *Limnology and Oceanography: Methods*, **6**, 144–152.

Index of Examples

General Index

Printed in the USA
CPSIA information can be obtained
at www.ICGtesting.com
JSHW060812260324
59916JS00006B/48